REVISE PEARSON EDEXCEL GCSE (9–1)

Statistics

REVISION GUIDE

Series Consultant: Harry Smith

Author: Su Nicholson

Also available to support your revision:

Revise GCSE Study Skills Guide 9781447967071

The **Revise GCSE Study Skills Guide** is full of tried-and-trusted hints and tips for how to learn more effectively. It gives you techniques to help you achieve your best – throughout your GCSE studies and beyond!

Revise GCSE Revision Planner 9781447967828

The **Revise GCSE Revision Planner** helps you to plan and organise your time, step-by-step, throughout your GCSE revision. Use this book and wall chart to mastermind your revision.

For the full range of Pearson revision titles across KS2, KS3, GCSE, Functional Skills, AS/A Level and BTEC visit:
www.pearsonschools.co.uk/revise

Contents

1-to-1 page match with the GCSE Statistics Revision Workbook ISBN 9781292191614

A small bit of small print
Edexcel publishes Sample Assessment Material and the Specification on its website. This is the official content and this book should be used in conjunction with it. The *Now Try This* questions have been written to help you practise every topic in the book.

Describing data

You need to be familiar with the terms for different types of data.

Raw data is data as it is first collected in a **statistical investigation**, before it has been sorted or ordered. Data is described in terms of the **variables** collected, which could be the colour of cars, engine size, number of people at a football match, length of time etc.

Raw data can be classed as either:

- ***quantitative*** – **numerical** data such as measures of height or weight
- ***qualitative*** – **non-numerical** data such as type of car or colour of hair.

Quantitative data

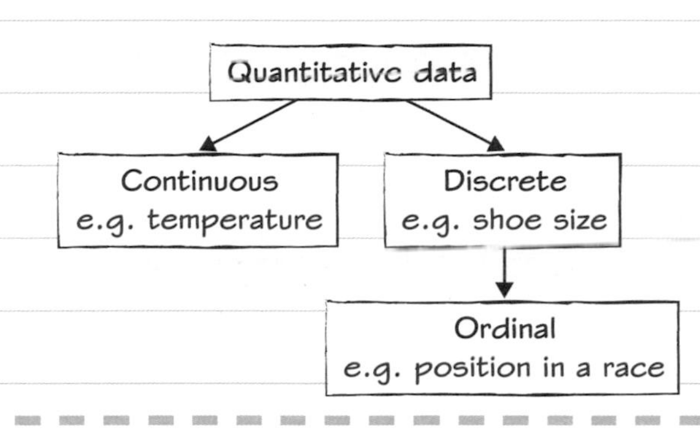

Qualitative data

Qualitative data are measures of types. Variables can be sorted into categories and are called **categorical data**. Categorical data is always qualitative. For example, raw data collected on where people went on holiday last year could be classed by continent: Europe, Asia, North America, South America, Africa, Australia and Antarctica.

Related data

Bivariate data involves **pairs** of related data values, such as exam results and time spent on study. **Multivariate data** involves sets of **three or more** related data values, such as age, height and weight.

Worked example

Here is a list of statistical words.

qualitiative	**quantitative**	**categorical**
bivariate	**ordinal**	**measurement**

Choose the word from the list that best describes the data below.

If none of the words is suitable then write 'none'.

(a) Number of cars in a car park **(1)**

Quantitative

(b) Shoe size **(1)**

Quantitative

(c) Position in a class test **(1)**

Ordinal

(d) Style of a painting **(1)**

Categorical

(e) Height and weight **(1)**

Bivariate

Positions in a class test are written in order of numerical outcome.

Look at the diagram above if you get stuck.

Now try this

tier F

1 Here are descriptions of four sets of data.

A Football scores
B Heights of plants
C Types of advert
D Positions of football teams in the Premiership

Which of these data sets are:

(a) ordinal **(1)**
(b) qualitative **(1)**
(c) quantitative continuous data **(1)**
(d) quantitative discrete data? **(1)**

2 Suggest a data item to complete a set of three multivariate data for a statistical investigation into people's choice of holiday destination.

Name of destination, temperature in month of holiday… **(1)**

Had a look ☐ Nearly there ☐ Nailed it! ☐

Primary and secondary data

Data can be categorised in two ways: as **primary** or **secondary** data.

Primary data

Primary data is information that you collect yourself.

You could do an experiment, carry out a survey or use a questionnaire to collect primary data.

Secondary data

Secondary data comes from published sources, such as newspapers, books or the internet.

You could take information from a table in a magazine to collect secondary data.

	Advantages	Disadvantages
Primary data	Collection method known. Accuracy known. Questionnaire or survey can be designed properly to find answers to specific questions.	Collection of data can be time-consuming and expensive.
Secondary data	Easy to obtain. Cheap to obtain. Data from known organisations (e.g. the UK Office for National Statistics) is usually reliable.	Data source may not be reliable. Data might contain errors. Data might not be suitable to find answers to specific questions. Collection method unknown (sample size, surveys or questionnaires might be misleading). Data might be out of date.

Worked example

Is the data collected in these examples primary data or secondary data? Give a reason why. **(4)**

(a) Karen investigates public opinion of the new leisure centre. She gets her data from people living in the local area.

This is primary data. Karen collects the information herself.

(b) A teacher wants to know if her students performed well in their GCSE Statistics exam. She looks at records of student results on the exam board website.

This is secondary data. The data is obtained from an organisation.

(c) Larry designs furniture. He finds published data on the internet for heights of people sitting at furniture to decide on the dimensions for a desk.

This is secondary data. Larry finds the information on the internet.

(d) Gav wants to decide which type of tomato plant to grow. He compares the weight of tomatoes he gets from five different types of tomato plant.

This is primary data. Gav collects the data himself.

Now try this

Jermaine gets a new job and is going to move to another city. He decides to investigate how safe different postcodes in the city are by looking at the Government crime statistics website.

(a) What type of data is this? Give a reason why. **(2)**

(b) How reliable is this data? Give reasons for your answer. **(2)**

(c) Give an example of a different type of data Jermaine could use to investigate how safe it is in a particular postcode area. **(1)**

Collecting data 1

Collecting data is very important in statistics. You need to know how to do it properly and how to criticise a method of data collection.

Data collection sheets

You can collect **primary data** through surveys, direct observations or experiments using a **data collection sheet.**

The data collection sheet should have three columns:

1. **subject** – the names of what you are collecting information on
2. **tally** – marks ~~||||~~ written as you count

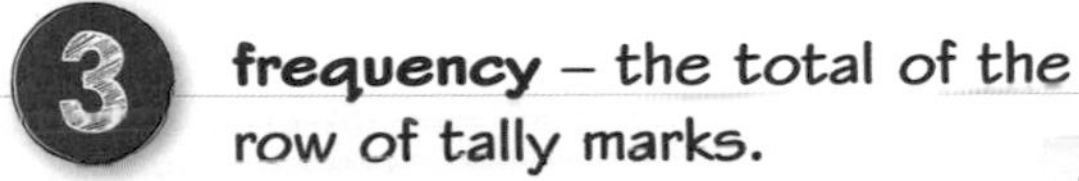

3. **frequency** – the total of the row of tally marks.

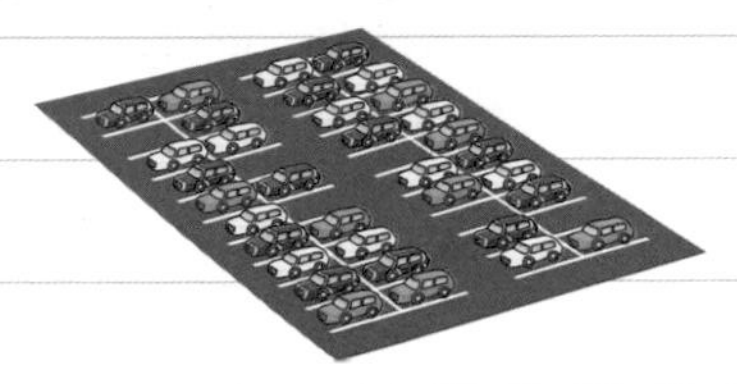

The first column lists all the car colours seen.

The Tally column shows the tallies. Group them in 5s.

Colour of car	Tally	Frequency
Blue	//	2
Green	~~////~~ //	7
Red	~~////~~ ~~////~~ ///	13
Yellow	~~////~~ ~~////~~ //	12

The Frequency column shows how many cars of each colour were seen in the car park.

Label the three columns.
Use a ruler to draw your table.
Add up the numbers in the frequency column and **check** against the total number of cars.

Worked example

tier F

Here is a list of the 20 trees in a wood.

Ash	Oak	Hazel	Ash	Beech
Hazel	Ash	Hazel	Ash	Beech
Hazel	Ash	Ash	Hazel	Hazel
Beech	Beech	Hazel	Oak	Hazel

Use a data collection sheet to record this information. **(3)**

Type of tree	Tally	Frequency
Ash	~~////~~ /	6
Oak	//	2
Hazel	~~////~~ ///	8
Beech	////	4

To design the data collection sheet:

1. Look through to see how many different types of trees there are.
2. Draw a three-column table with headings for Type of tree, Tally and Frequency.
3. Write the names of the different types of trees in the first column.
4. Use the Tally column to record the number of each type of tree. Cross each tree off the list as you tally it.
5. Add up the tallies for each row and write the totals in the Frequency column.

Check it!
$6 + 2 + 8 + 4 = 20$ ✓

Now try this

tier F

1 Jessie wants to find out the number of men, the number of women, the number of boys and the number of girls using a sports centre one night. Design a data collection sheet that she could use. **(3)**

2 Here is a list of some precious stones.

Ruby	Diamond	Emerald	Ruby	Sapphire	Ruby
Emerald	Ruby	Diamond	Emerald	Sapphire	Sapphire
Ruby	Emerald	Ruby	Diamond	Diamond	Sapphire
Emerald	Emerald	Ruby	Diamond	Ruby	Sapphire

Record this information in a data collection sheet. **(3)**

Had a look ☐ Nearly there ☐ Nailed it! ☐

Collecting data 2

Researchers commonly collect data from different types of **experiment.**

In each type of experiment, the researcher is interested in the ways that changes in one variable (the **explanatory** or **independent** variable) affect another (the **response** or **dependent** variable).

Extraneous variables need to be controlled. These are any variables the researcher is not interested in but could affect the results of the experiment.

When **replicating** or repeating an experiment gives very similar data, it is likely that the data is valid and reliable.

Controlled experiments

Laboratory experiments are conducted in a **controlled** environment.

Example: To investigate the effect of sunlight on plant colour, plants are exposed to sunlight for varying lengths of time.

Explanatory variable: length of time for sunlight

Response variable: difference in colour of plant

Advantages: easy to replicate; can control extraneous variables such as changes in the weather

Disadvantages: plants may react differently in a controlled environment compared with real-life growing conditions

Worked example

tier F&H

Sylvie wants to investigate whether a new type of medication helps people who have difficulty sleeping. She plans to run a laboratory experiment in a hospital with patients who suffer from this condition.

(a) Identify the explanatory and response variables in this experiment. **(2)**

The explanatory variable is the type of medication.

The response variable is the degree of difficulty the patient has in sleeping.

(b) Describe one advantage and one disadvantage of doing this as a laboratory experiment. **(2)**

An advantage is the medication can be controlled and the amount of sleep can be measured.

A disadvantage is that the patient may experience difficulty in sleeping because they are out of their home environment.

This is also an extraneous variable which is difficult to control in this laboratory experiment.

Now try this

tier F&H

Brad wants to investigate whether a dog's coat improves if it is fed with a new formula of dog food. He plans to run a laboratory experiment with a sample of pet dogs.

(a) Identify the explanatory and response variables in this experiment. **(2)**

(b) Describe one advantage and one disadvantage of doing this as a laboratory experiment. **(2)**

Collecting data 3

Experiments can also be carried out in **uncontrolled environments.**

In the same way as for controlled experiments, the researcher will look at the ways that changes in the **explanatory** or **independent** variable affect the **response** or **dependent** variable. Also, any **extraneous variables** need to be identified and controlled.

Uncontrolled experiments

Field experiments are carried out in an everyday (uncontrolled) environment. The researcher sets up the situation and variables are controlled.

Example: To investigate a new treatment for arthritis, the mobility of a sample of people is assessed before the treatment and then again after the treatment.

Explanatory variable: new treatment

Response variable: difference in mobility

Advantages: more likely to reflect real-life behaviour

Disadvantages: cannot control extraneous variables, e.g. how active a person is at home and what effect that might have on mobility

Natural experiments are carried out in an everyday (uncontrolled) environment. The researcher has no control over any variables.

Example: To investigate how a reduction in the speed limit affects the number of car accidents on a road, the number of accidents per year on the road before the reduction is compared with the number per year after the reduction.

Explanatory variable: speed limit

Response variable: number of accidents

Advantages: more likely to reflect real-life behaviour

Disadvantages: cannot control any variables; harder to replicate the study

Worked example

tier F&H

Claire investigates whether having breakfast in the morning helps student performance.
On Day 1, she asks her students to come into college without having breakfast before they take a maths test.
On Day 2, they are asked to have breakfast before coming into college to take another similar maths test.

(a) What type of experiment is this? **(1)**

A field experiment.

(b) Give one advantage and one disadvantage of doing this type of experiment. **(2)**

Advantage: more likely to reflect real-life behaviour

Disadvantage: cannot control extraneous variables, such as amount of sleep, which may affect performance

(c) Identify one possible extraneous variable. **(1)**

Student performance in second test may improve if they learn from the first test.

Now try this

tier F&H

Fred wants to investigate whether the amount of rainfall in the summer affects the number of visitors to a museum. He plans to record the total rainfall in mm and the number of visitors to the museum during the summer months for successive years.

(a) What type of experiment is this? **(1)**

(b) Explain what is meant by the term explanatory variable. **(1)**

(c) Identify one possible extraneous variable. **(1)**

Problems with collected data

You need to know the **problems** that can occur with **collected data** and how to deal with them.

Checking data

It is important to check collected data before you process it, to ensure it is consistent and accurate, otherwise your results may be invalid. Collected data may contain **outliers** or **anomalous values** that do not fit the pattern of the rest of the data.

Outliers can be ignored if they are due to measuring or recording error.

You also need to consider how your data has been collected (your **collection plan**) to see if this could affect the reliability of your results.

Cleaning data

If you find problems with your data it must be **cleaned**.

To clean data you must

- identify and correct or remove inaccurate data values or extreme values
- check units are consistent
- record values without units or other symbols
- decide what to do about missing data.

Worked example

tier F&H

In a medical study, adults were asked to record their height and weight. Here is some of the data.

	1	2	3	4	5	6	7	8	9
Height	162 cm	5 ft 7 in	1.72 m	166 m	6 ft 1 inch	158 cm	1.43 m	1.5 m	5 ft
Weight	110 kg	16 stone	10 kg		130 kg	116 kg	10 stone 3 pounds	105 kg	8 stone

(a) Give **two** reasons why this data must be cleaned before it is processed. **(2)**

The data values are in different units.
There is one data value missing.

(b) Explain with reasons which values are likely to have been written incorrectly. **(2)**

166 m is more likely to be 166 cm because the height of a person cannot be 166 m.
10 kg is an impossible weight for an adult.

(c) Discuss how this collection plan could affect the reliability of the conclusions. **(2)**

This is a small sample on which to base conclusions.
There are issues with data collection because people are recording their own data. This might decrease reliability.

Now try this

tier F&H

Brad uses a memory test to investigate whether having 8 or more hours' sleep helps student memory. Students are given a list of 20 words to look at for 30 seconds, then asked to write down the words they can remember. On Day 1, they take the test after having less than 8 hours' sleep. On Day 2, they take the test after having 8 hours or more sleep.

These are the results for the 75 students in the experiment:

- On Day 2, 74 students remembered more than 15 words.
- Of those 74 students, 70 students remembered 2 or more words **more** on Day 2 than they did on Day 1.
- One student remembered 14 words on Day 1 but only 8 on Day 2.

Discuss whether or not to include this student's results. **(2)**

Populations

Technical words and phrases

You need to know these definitions:

A **population** is everything or everybody that could possibly be involved in an investigation, e.g. students in a school, all the people who use the local gym.

A **census** gathers data from the whole population.

A **sample** gathers data from some of the population.

A **representative sample** should contain all the characteristics of the population to avoid **bias**. A sample that is too small may not represent the population and may bias the results.

The **sampling units** are the people or items that are to be sampled.

A **sampling frame** is a list of all the members of the population from which the sample will be taken.

A **pilot survey** is a small sample analysed first before any large-scale samples.

A **pre-test** is a pilot where questions for a questionnaire are usually tried out.

Census vs sample

Here is a population.

A census would gather information from **everyone**.

All the members of the population can be numbered to form a **sampling frame**.

A **representative sample** is shown in red.

A census collects more information than a sample but takes a lot longer and is a lot more expensive.

Good and poor samples

Good samples	Poor samples
are as large as possible	are too small
are representative	are biased – they unfairly favour one set of values
have a suitable sampling frame	have a poor sampling frame (e.g. out of date, people missing, people counted twice, names on a list that shouldn't be there)

Worked example

The manager of a new leisure centre wants to find out what people who live in the local area think about the quality of the facilities in the leisure centre.

(a) Write the population the manager should use. **(1)**

The population is all the people in the local area who use the leisure centre.

(b) Describe a sampling unit. **(1)**

A person who lives in the local area and uses the leisure centre.

The manager asks 20 people who come to the leisure centre on Monday morning.

(c) Give two reasons why this sample is likely to be biased. **(2)**

The sample is not representative of all the people who use the leisure centre at different times and may include people who do not live in the local area.
The sample is too small.

The population will not be everybody in the local area, because the manager wants to know about the **users** of the leisure centre.

The sample should represent the whole population and include people who use the leisure centre at different times of day and week. A sample that is too small can bias the results. In general, the larger the sample the more reliable the results.

Now try this

Jim wants to find out how many of the 250 students in his year bring a mobile phone to school. He decides to ask 10 of his friends.

(a) Write down two reasons why this is not a good sample. **(2)**

(b) Explain how Jim could take a better sample. **(1)**

Grouping data

Data can be **grouped** to help you see the distribution of the data and identify patterns.

You need to know these terms:

- **Class intervals** are groups that do not overlap.
- **Upper and lower class boundaries** are the boundaries between one class and the next.
- **Class width** is the difference between the upper class boundary and the lower class boundary.

Worked example

The heights, h cm, of 30 female students correct to the nearest 0.1 cm are shown here.

151.2	156.3	160.1	165.8	149.5
150.1	161.6	174.4	173.2	152.3
160.4	171.8	157.2	173.9	156.8
166.4	160.4	159.2	147.9	166.2
164.1	166.8	170.0	169.2	157.8
149.2	164.7	174.1	155.5	167.6

(a) Suggest suitable class intervals for this data. **(2)**

Class intervals
$145 \leqslant h < 150$
$150 \leqslant h < 155$
$155 \leqslant h < 160$
$160 \leqslant h < 170$
$170 \leqslant h < 175$

A student who is taller than the other students joins the group.

(b) How could you change your final interval to allow for taller people? **(1)**

Add an extra interval, $h \geqslant 175$

You should use as few groups as is reasonable. If you have too few, or too many, the pattern of the data may not be clear. You can combine adjacent groups if they contain only a few data values.

The number of class intervals and the width of the intervals can help you spot patterns in the data.

The first interval contains all the values from 145 up to but not including 150. You do not need to start with the smallest data value of 147.9.

For continuous data, there must not be any gaps between the intervals. Use inequalities so that values up to but not including 155 are in the second interval. The third interval starts with 155.

The height 170.0 cm belongs in the last interval.

Intervals do not have to be of equal width. Add an open interval for taller people.

Now try this

tier F

Twenty new-born babies were weighed. These are their weights to the nearest 0.01 kg.

3.12	3.90	2.95	3.08	4.13
4.01	3.76	4.44	3.26	2.93
3.62	4.07	3.49	4.50	2.99
4.18	3.81	4.09	3.28	4.80

Comment on the suitability of each of these sets of class intervals.

(a)

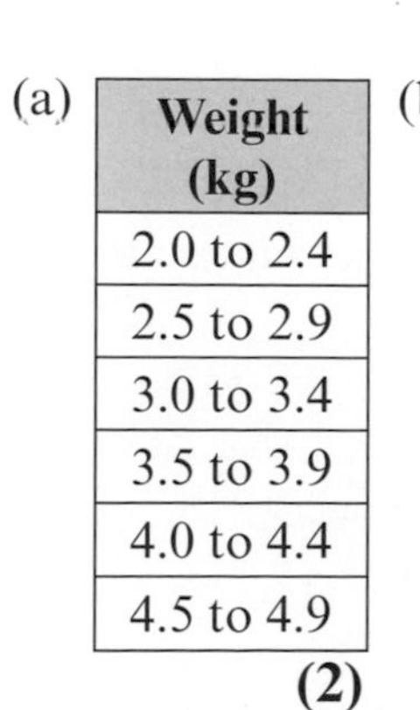

Weight (kg)
2.0 to 2.4
2.5 to 2.9
3.0 to 3.4
3.5 to 3.9
4.0 to 4.4
4.5 to 4.9

(2)

(b)

Weight (kg)
2.0 to 2.5
2.5 to 3.0
3.0 to 3.5
3.5 to 4.0
4.0 to 4.5
4.5 to 5.0

(2)

(c)

Weight, w (kg)
$2.0 < w \leqslant 2.5$
$2.5 < w \leqslant 3.0$
$3.0 < w \leqslant 3.5$
$3.5 < w \leqslant 4.0$
$4.0 < w \leqslant 4.5$
$4.5 < w \leqslant 5.0$

(2)

Random sampling

A **random sample** is one in which every member of the population has an **equal chance** of being selected. A random sample is fair or **unbiased** and, if it is large enough, it is more likely to be **representative** of the population.

Methods for random sampling

Give each item in your sampling frame a unique number. To select the numbers for the items in your sample, you can:

- use a random number table
- use a random number generator on a computer or calculator
- put the numbers of the items on pieces of paper and select at random from a hat
- roll sets of fair 10-sided dice to generate digits from 0 to 9.

Advantages and disadvantages of random sampling

Advantages	Disadvantages
Random sample is more likely to be representative of the population, provided the sample is large enough.	A full list of the whole population is needed.
Choice of members of sample is unbiased.	A large sample size is needed.

Worked example

tier F&H

(a) Explain what is meant by a random sample. **(1)**

A random sample is one in which every member of the population has the same chance of being selected.

You need to learn this definition. Don't use 'random' or 'no pattern' in your explanation.

(b) Amina is going to take a random sample of 200 houses from the 4000 houses in her town. Explain how Amina could select a random sample. **(3)**

Write a sampling frame by listing the street names and the house numbers in order.

Give each house a different number from 1 to 4000.

Generate 200 random numbers between 1 and 4000 then use these numbers to pick the houses in the sample, e.g. 2160, 572, 1708, 97, 220, 8 ...

To generate a random whole number between 1 and 4000 use the RanInt button on your calculator to key in RanInt # (1, 4000).

Now try this

tier F&H

Karl needs to select a random sample from a numbered list of 750 people.

He uses his calculator to generate these random numbers in decimal form.

0.583	0.958	0.196	0.811
0.68	0.043	0.326	0.374
0.416	0.006	0.334	0.719

(a) Explain how Karl can use these numbers to select a sample of 10 people from the list of 750 people. **(3)**

(b) Comment on the reliability of his sample. **(1)**

Had a look ☐ Nearly there ☐ Nailed it! ☐

Stratified sampling 1

Stratified sampling can be used when the population can be split into distinct **groups**. The advantage over simple random sampling is that you can be certain that all the groups are represented in the sample. This may not be true for simple random sampling.

Strata

A **stratum** is a group in the population. In a stratified sample, the relative sizes of the groups in the sample are the same as their relative sizes in the whole population.

There are twice as many boys as girls in this population.

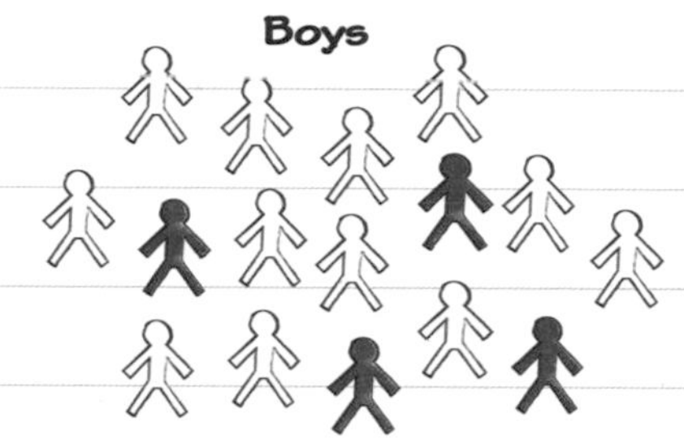

So you need twice as many boys as girls in the stratified sample (shown in red).

Worked example

This table shows the numbers of males and females who belong to a gym.

Males	Females	Total
120	80	200

A sample of 40 gym members stratified by gender is to be taken. Work out the number of males in the sample. **(2)**

Proportion of males in population

$= \frac{120}{200} = 0.6$

Number of males in sample

$= 0.6 \times 40 = 24$

The proportion of males in the sample must be the same as the proportion of males in the population. To select this sample the sampling frame must contain data about the gender of each member of the population.

Worked example

A primary school has three classes in Year 6.

Class	A	B	C
Number of children	34	28	18

The headteacher wants to take a sample of 20 children stratified by class.

(a) What is the advantage of using a stratified sample? **(1)**

Each class is represented in the sample in proportion to its size.

(b) Work out the number of children in the sample for each class. **(3)**

There are $34 + 28 + 18 = 80$ children in Year 6.

Class A: $\frac{34}{80} \times 20 = 8.5$

Class B: $\frac{28}{80} \times 20 = 7$

Class C: $\frac{18}{80} \times 20 = 4.5$

$8.5 + 7 + 4.5 = 20$

Class	A	B	C
Number in sample	9	7	4

Number in the sample for each stratum

$= \frac{\text{stratum size}}{\text{population size}} \times \text{number in sample}$

There must be a whole number of children for each class, so round to the nearest integer. Rounding gives a total of 21, so subtract 1 from the smallest stratum, i.e. decrease 5 to 4.

Now try this

There are 520 young people in a youth club. 280 are boys.
The youth club leader wants to take a stratified sample of 50 young people.
Work out how many boys should be in the sample. **(2)**

Non-random sampling

You need to know about these methods of **non-random sampling.**

1. **Judgement sampling** uses judgement to select a sample that is representative of the population.
2. **Opportunity sampling** uses the people (or objects) that are available at the time.
3. **Cluster sampling** can be used when the population is in groups. A random sample of these groups is selected and all items in the selected groups are included in the sample.
4. **Quota sampling** involves splitting the population into groups with certain characteristics (e.g. age, gender) and selecting a given number from each group. For example, a market researcher might ask 10 adults and 10 children about their reaction to the 2018 GCSE results.
5. In **systematic sampling**, items are selected from the population at regular intervals either in time or in space. For example, every 5th car that passes a location or every 3rd house on a street.

Worked example

tier F&H

(a) Describe the difference between an opportunity sample and a cluster sample. **(1)**

Opportunity sample: uses the people who are available at the time.

Cluster sample: the clusters are selected at random and then all the items in the selected clusters are included.

A city has 250 dental surgeries employing over 600 dentists in total. A researcher wants to carry out face-to-face interviews with a sample of 70 dentists.

(b) Explain why a cluster sample would be a suitable way of carrying out the interviews. **(1)**

The surgeries are likely to be geographically spread out, so it is more efficient to interview all the dentists at a small number of surgeries.

(c) Describe what the sampling frame would be in this case. **(1)**

A list of all the dental surgeries in the city, written in alphabetical order or postcode order. This would make it easy to take a random sample.

You must answer in context and not just give a general definition of a sampling frame. Here you must refer to a list of surgeries and how they are ordered.

Now try this

tier H

1 A street contains 160 houses. There are 80 houses on each side of the street, arranged in 20 blocks of 4 houses each. A researcher wants to take a systematic sample of 40 houses from the street.

(a) Describe a simple way in which he could do this. **(1)**

(b) Describe one disadvantage of this. **(1)**

(c) Describe how to re-design the systematic sample to reduce the impact of the disadvantage you have described in part (b). **(2)**

2 About 70% of expensive women's perfume is bought by women for themselves and about 30% is bought by men for women.
A marketing company wants to test out people's opinion of a new perfume. It has enough money to interview 40 people and must get the results quickly.

(a) Explain why quota sampling may be a suitable method. **(1)**

(b) Describe how this could be carried out. **(2)**

Had a look ☐ Nearly there ☐ Nailed it! ☐

Stratified sampling 2

Stratified sampling using two-way tables

You can use each cell in a **two-way table** as a stratum.

You need to ensure that the proportion of the stratum in the sample is the same as its proportion in the population.

The table shows the numbers of female and male car drivers and HGV drivers.

See page 19 for more about two-way tables.

One stratum (one of the cells) is that of female car drivers.

	Female	Male
Car drivers	1489	1876
HGV drivers	238	3845

The proportion of this stratum in the population is $\frac{1489}{1489 + 1876 + 238 + 3845} = \frac{1489}{7448}$

So the number of female car drivers in the sample should be $\frac{1489}{7448} \times$ the size of the total sample.

Worked example

tier F&H

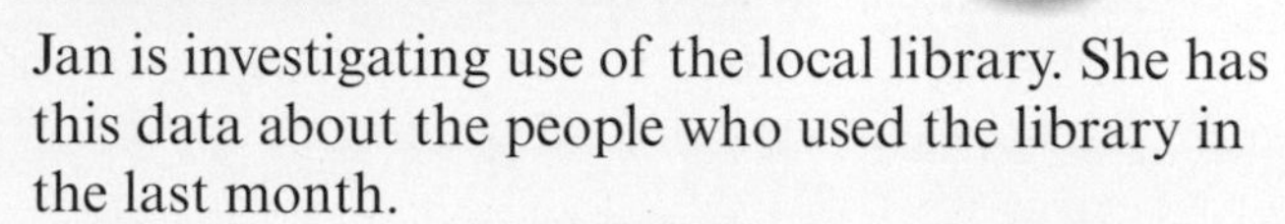

Jan is investigating use of the local library. She has this data about the people who used the library in the last month.

	Female	Male
Under 25	1543	878
25 and over	5060	3789

Jan wants to take a sample of 120 library users stratified by age and gender.

(a) Explain why a stratified sample is a good choice of data collection for her investigation. **(2)**

This is a suitable choice for data collection as it will produce a representative sample of the people who use the library.

Calculate the proportion for each stratum and multiply by the population size for each stratum.

Golden rule

If your sample size is too small you need to add 1 to the largest stratum.

(b) Complete this table to show how many people in each of the four groups should be in the sample. **(4)**

	Female	Male
Under 25	16	9
25 and over	~~54~~ 55	40

Work out the total population size.

Total population size is:

$1543 + 5060 + 878 + 3789 = 11270$

Proportion of under-25 females $= \frac{1543}{11270}$

$\frac{1543}{11270} \times 120 = 16.42$, rounded to 16

Proportion of 25+ females $= \frac{5060}{11270}$

$\frac{5060}{11270} \times 120 = 53.87$, rounded to 54

Round to the nearest whole number. Check that your total sample is the correct size. $16 + 9 + 54 + 40 = 119$ so add 1 to the stratum of females 25 and over.

Now try this

tier H

Here are the numbers of patients attending a surgery.

	Male	Female
Adult	48	64
Child	28	30

The doctor wants to take a sample of 50 patients stratified by age and gender. Work out the number of male adults there should be in the sample. **(3)**

Petersen capture-recapture formula

You can use the **Petersen capture-recapture formula** to estimate population size for large populations. The diagrams below show how this method works on a population of fish.

Estimating populations

Catch a sample of fish and mark them. Record how many have been marked and return them.

The experimenter catches 4 fish, marks them and returns them.

2 Catch a second sample of fish and count how many in this sample are marked.

The experimenter catches 5 fish in a second sample. 2 are found to be marked.

Using ratios

You can compare **ratios** to estimate the size of the population. You **assume** that these two ratios are equivalent:

Ratio of marked fish : total fish in recapture sample

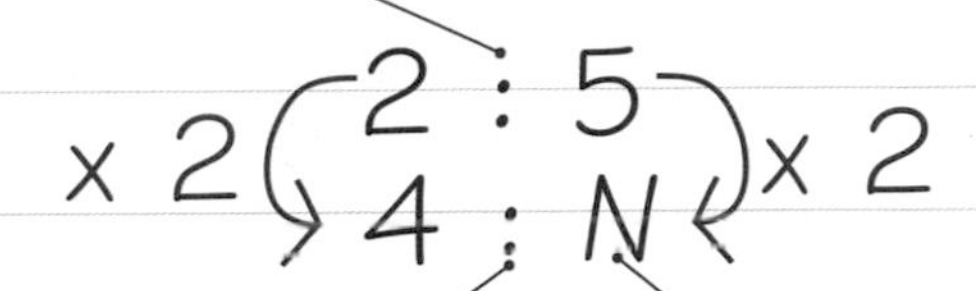

Ratio of marked fish : total fish in whole population

N is the population size

Using the formula

You can use this formula to estimate the population size, N:

$N = \frac{Mn}{m}$ or $\frac{m}{n} = \frac{M}{N}$ **LEARN IT!**

M = number of fish marked then released

n = size of recapture sample

m = number of marked fish in recapture sample

Worked example

A scientist wants to estimate the number of mice on an island. She captures 60 mice, marks them and releases them. Later she captures 40 mice, of which 6 were already marked.

(a) Find an estimate for the number of mice on the island. **(2)**

If the number of mice on the island is N, then

$\frac{N}{60} = \frac{40}{6}$ so $N = \frac{60 \times 40}{6} = 400$

(b) Write down two assumptions the scientist has to make. **(2)**

1. The marked mice do not die.
2. The probability of being caught is the same for marked and unmarked mice.

Assumptions

Learn the underlying assumptions.

✓ The population is closed – no migration.

✓ All members of the population are equally likely to be captured in each sample.

✓ Capture and marking do not affect catchability and markings are not lost.

✓ The population does not change due to deaths or births between sampling occasions.

✓ The sample is large enough to be representative of the population.

Now try this

tier H

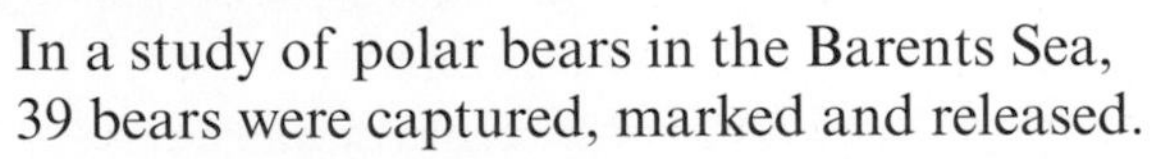

In a study of polar bears in the Barents Sea, 39 bears were captured, marked and released.

One year later a sample of 120 bears yielded 9 that were marked.

(a) Work out an estimate for the number of polar bears in the Barents Sea. **(2)**

(b) Write down one assumption that has to be made. **(1)**

Had a look ☐ Nearly there ☐ Nailed it! ☐

Controlling extraneous variables 1

Extraneous variables are variables that you are not interested in as part of an investigation or experiment but that could affect your results. For example, in an investigation into whether background music increases production in a factory, the health or tiredness of the employees may affect the results.

You need to know the importance of identifying and controlling extraneous variables, so that your results only focus on the connection between the variables you are interested in.

Worked example

tier F&H

In the memory test experiment to remember a list of words described on page 6, students took the test on Day 1 in the main hall at school.

On Day 2, the test was also taken in the main hall but there was building work going on outside.

(a) What is the explanatory variable in this experiment? **(1)**

The explanatory variable is the list of words.

(b) What is the response variable in this experiment? **(1)**

The response variable is the number of words remembered.

(c) Identify an extraneous variable in this experiment and describe how this could be controlled. **(2)**

An extraneous variable is the noise from the building work. This could be controlled by the students using noise-cancelling headphones when taking the tests.

Look at page 4 for a reminder about explanatory and response variables.

Extraneous variables could include any source of distraction such as noise, behaviour, lighting and temperature.

Examples of extraneous variables	How to control
Noise and distractions e.g. building work outside the room where an experiment is taking place.	Make use of noise cancelling headphones.
Temperature e.g. no air conditioning in the room on a hot day.	Make use of fans, open doors and windows.
Location e.g. experiment requiring an overnight stay in unfamiliar surroundings.	Consider conducting the experiment in the participant's own home and ask them to feed back with outcome.
Lighting e.g. poor lighting in a room where experiment requires identifying words or objects.	Introduce additional lighting making sure all participants can see well enough to take part in the same conditions.

Now try this

tier F&H

A researcher carries out an investigation into a new medication to aid sleep. Patients are brought into hospital and their heart rates are monitored through the night.

(a) What is the explanatory variable in this experiment? **(1)**

(b) What is the response variable in this experiment? **(1)**

(c) Identify **one** extraneous variable and describe how the researcher could control it. **(2)**

Controlling extraneous variables 2

Researchers will often make use of **control groups** or **matched pair tests** in their investigations. This can reduce the effect of extraneous variables.

Control groups

A **control group** is used to test the effectiveness of a treatment. Random selection is used to select two groups of people. In an experiment to test a new drug, a test group is given the treatment and the control group is given no treatment or an inactive treatment (called a **placebo**). Members of each group do not know whether they receive the active or inactive treatment so the results can be compared to see how effective the treatment is.

Matched pair tests

Matched pair tests can be used to test the effects of a particular factor. Each individual in one group is paired with an individual in the second group. The two individuals have everything in common apart from the factor being studied.

Worked example

tier H

A new tablet for arthritis is being tested. One group of patients is given the medication. The control group is given tablets which look identical but do not contain any medication. The severity of arthritis is measured on a scale from 1 to 100.

(a) Explain how matched pairs could be used for this investigation. **(2)**

Think about the extraneous variables. How can you match pairs to control these?

Patients could be paired by age, gender and level of severity. For example:

	Group 1 (medication)	Group 2 (no medication)
Pair A	Male, age less than 30, level less than 15	Male, age less than 30, level less than 15
Pair B	Female, age 30 to 40, level 15 to 25	Female, age 30 to 40, level 15 to 25

Suggest some matched pairs to use.

(b) Describe **one** advantage and **one** disadvantage of using matched pairs in this example. **(2)**

An advantage of using matched pairs is that effects of gender, age or different levels of severity can be controlled.

A disadvantage is that it can take a lot of time to find enough matched pairs for a good test.

Now try this

A gardener wants to know whether a new fertiliser has an effect on the growth of his rose bushes. He divides the rose bushes into two groups and gives one group the fertiliser. The control group does not get any fertiliser.

(a) Explain the purpose of using a control group. **(1)**

(b) Identify **one** extraneous variable and describe how the gardener could control it. **(2)**

Had a look ☐ Nearly there ☐ Nailed it! ☐

Questionnaires and interviews 1

A **questionnaire** is a set of questions designed to collect primary data.
The person who completes the questionnaire is called the **respondent**.
Questionnaires are given to people to complete anonymously, either printed or online.
Interviews are usually carried out in person or by telephone.

	Advantages	Disadvantages
Questionnaires	• Much cheaper to do • Each person answering the question is treated in the same way	• Can be inflexible • People may misunderstand some questions
Interviews	• Interviewer can explain complex questions • Interviewer can follow up on unclear responses	• Interviewer may be biased • Can be costly

Types of questions

Avoid **open** questions which allow a wide variety of responses.

e.g. 'What do you think about programmes on TV?'

Use **closed** questions to restrict the replies given.

e.g. 'Are you over 18 years old?'

Avoid **leading or biased questions** which might **lead** the respondent towards the answer that you want or expect.

Golden rules

Remember these rules for designing questionnaires:

- ✓ Make questions clear and closed.
- ✓ Avoid open questions.
- ✓ Don't ask leading questions.
- ✓ Have response boxes which are unambiguous.
- ✓ Have response boxes which cover all possible replies and don't overlap.

Worked example

tier F&H

Akbar wants to know how far people will travel to buy organic food. He asks people this question on a questionnaire:

How far do you travel to buy organic food?

[1–2 km] [3–5 km] [6+ km]

(a) Write down three things that are wrong with this question. **(3)**

1. No time frame
2. No distance less than 1 km in the response boxes
3. Gaps in the distances (e.g. 2.5 can't be placed)

When asked to criticise a question on a questionnaire, ask these questions:

- Is there a time frame?
- Do the response boxes overlap?
- Do the response boxes cover all possible responses?

Now try this

tier F&H

Mr Brown owns a café. He wants to find out what people think of the service in the café.
He uses this question in a questionnaire:

What do you think of the service in the café?

☐ *excellent* ☐ *very good* ☐ *good*

You can comment on either the question or the response boxes.

(a) Write down one thing that is wrong with this question. **(1)**

(b) Write an improved question. You must include response boxes. **(2)**

Questionnaires and interviews 2

Pilot surveys

Pilot surveys are used to test questions in a questionnaire, to make sure that respondents understand the questions and can answer in ways that will collect all the data needed and give valid results. Pilot surveys are usually carried out on a proportion of the total sample population.

Worked example

Abbi wants to use a questionnaire to investigate insurance claims.

(a) State **two** reasons why Abbi should carry out a pilot survey. **(2)**

1. To see if there are any problems with the wording or the response boxes.
2. To make sure the questionnaire will collect the information needed.

This is a question on Abbi's questionnaire:

> Roll a fair dice. If you get a 6, tick box A.
> If you get 1, 2, 3, 4 or 5, answer this question truthfully:
> *Have you ever broken something and not owned up to it?*
> *If yes, tick box A ☐ If no, tick box B ☐*

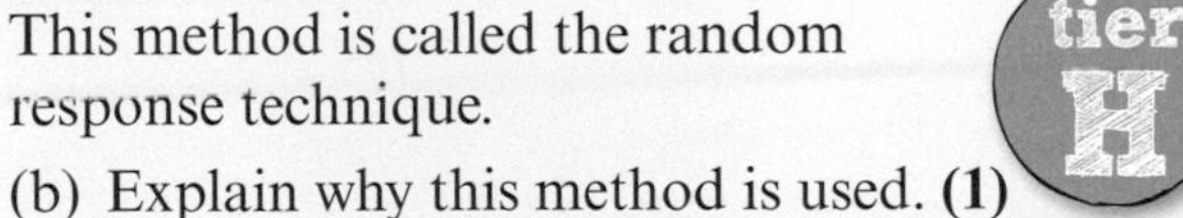

This method is called the random response technique.

(b) Explain why this method is used. **(1)**

This is a sensitive question and people may not want to answer it otherwise.

432 people ticked box A.
468 people ticked box B.

(c) Calculate an estimate for the proportion of the people in the sample who had broken something and not owned up to it. **(3)**

Work out the total number of people who answered the question.

$432 + 468 = 900$

Calculate an estimate for the number of people who roll a 6.

$P(\text{roll a }6) = \frac{1}{6}$

Estimated number of 6s $= \frac{1}{6} \times 900 = 150$

Subtract the estimated number of people who rolled a 6 and ticked box A from the total number who ticked box A.

Estimate for the number who ticked box A that were truthful $= 432 - 150 = 282$

Estimated proportion of people who had broken something and not owned up to it

$= \frac{282}{900 - 150} = \frac{282}{750} = 0.376$

$\frac{282}{\text{number of people who answered truthfully}}$

Random response method

People do not always answer questions truthfully, especially if a questionnaire asks sensitive questions. This means results may not be accurate.

The **random response** method uses a random event, such as tossing a coin, to obtain truthful answers.

For example, if you asked 50 students:

> Toss a coin. If it lands on heads, tick 'Yes' below.
> If tails, answer the question truthfully.
> *Have you ever been bullied? Yes ☐ No ☐*

you would expect 25 students out of 50 to have tossed heads, so they will tick Yes. If your results show that 34 have ticked Yes, then about 9 of the students who ticked Yes have been bullied. So you can estimate that 18 of the 50 have been bullied.

Now try this

This question is given to a sample of people.

Have you ever pretended to be someone else?

Flip a coin: If you get heads, answer 'Yes'. If you get tails, answer truthfully.

520 people answered Yes and 480 people answered No.

Estimate the proportion of people in the sample who had pretended to be someone else. **(3)**

Had a look ☐ Nearly there ☐ Nailed it! ☐

Hypotheses

A **hypothesis** is a precise statement about something you can measure that can be tested by collecting and analysing data. It may or may not be true. You need to collect relevant data and analyse it to see whether it supports or contradicts your hypothesis.

A hypothesis must be

- **specific** e.g. 'swimming for 5 hours or more a week improves your heart rate' cannot be misinterpreted.
- **measurable** e.g. 'as people get older they need more sleep' can be analysed by recording age and amount of sleep as grouped data, using age in years and amount of sleep in hours.

Worked example

tier F

Matt writes this hypothesis:
Young people spend more time at the gym than old people.

(a) Explain why this is not a good hypothesis. **(1)**

The statement is not precise and not measurable. 'Young' and 'old' are not defined.

(b) Write a better hypothesis Matt could use. **(1)**

People under 30 spend more time at the gym than people over 50.

When you write your hypothesis you need to think how the data can be recorded for ease of collection and analysis. Here you can use these age groups:

- less than 30 years
- 30 to 50 years
- over 50 years

The timeframe for the data collection also needs to be identified. For example:
How many hours do you spend at the gym on average per week?

- 0
- fewer than 5
- 5 to 10
- more than 10

Worked example

Rosa wants to investigate the amount of time male and female students spend using their mobile phones each day.

(a) Write a hypothesis Rosa could use. **(1)**

Female students spend more time on their mobile phones than male students.

(b) Explain what data she should collect to test the hypothesis, stating whether it is primary or secondary data. **(2)**

Data on amount of time spent per day on mobile phones (to nearest minute) by males and females. This is primary data.

Other answers could be:

- Males spend more time on their mobile phones than females.
- Females spend the same amount of time on their mobile phones as males.

Now try this

Kyle wants to investigate how people do their clothes shopping: online, at local shops or at large retail outlets.

(a) Write a hypothesis he could use. **(1)**

(b) Describe how he could collect primary data to test his hypothesis. **(1)**

(c) He wants to distribute a sheet to people in the local area to record how they do their clothes shopping. Design a suitable data collection sheet for this data. **(2)**

Designing investigations

There are **constraints** which must be considered when **designing an investigation** to test a hypothesis.

Considerations

- **Time** and **cost** to set up and carry out investigation.
- **Ethical issues** You must respect people's dignity and rights.
- **Confidentiality** It is important to keep data secure and confidential.
- **Convenience** of getting data locally.
- **Identifying the population** and method to collect sample data.
- Planning to gain more responses than you think you need, in case of **non-response.**
- Doing a **pilot survey** to help work out likely responses to sensitive questions.
- Planning what to do with **anomalous results.** (Read page 5 for a reminder about anomalous results.)

Planning for non-response

- Decide on the number of responses you need to do a valid analysis of data.
- Do a pilot survey to work out the proportion of surveys which are likely to be returned.
- Use this proportion to work out how many surveys to send.

Calculating the number of surveys

For a survey, 300 responses are needed. In the pilot survey, 50 questionnaires are sent out. 40 responses are received.

The proportion of responses received is $\frac{40}{50} = \frac{4}{5}$

The total number of surveys to be sent out can be called x, so to receive 300 responses to the full survey, you need:

$\frac{4}{5} = \frac{300}{x}$, so $x = \frac{5 \times 300}{4} = 375$ surveys

You need to send out 375 surveys.

Worked example

Zeedan wants to investigate whether people in the UK prefer to drink tea or coffee. He sends out a pilot survey to 270 people and gets 180 completed surveys back.

(a) Zeedan wants to get at least 400 completed surveys. How many people should he send the survey to? **(2)**

Using proportion:

$\frac{180}{270} = \frac{400}{x}$ so $x = \frac{400 \times 270}{180} = 600$

Zeedan should send the survey to at least 600 people.

Let the number of surveys Zeedan needs to send out be x. Check your answer – x must be greater than 400 which is the number of completed surveys he needs.

Now try this

tier F&H

An insurance company wants to investigate how many times people have claimed on their insurance for theft. They realise that this could be a sensitive issue.

(a) Discuss whether they should use an interview or a questionnaire. **(2)**

(b) Some data for insurance claims can be obtained from the internet. Give **one** advantage and **one** disadvantage of using data from the internet. **(2)**

This could be a sensitive issue as people may not like to admit to being victims of theft.

Had a look ☐ Nearly there ☐ Nailed it! ☐

Tables

When extracting information from tables make sure you read the table and the units carefully.
Figures in tables can sometimes be **rounded**.
This table shows the numbers of people who voted in three General Elections.

	Voters (millions)					
Year	**CON**	**LAB**	**LD**	**PC/SNP**	**Other**	**Total**
2010	10.70	8.61	6.84	0.66	2.88	29.69
2015	11.30	9.35	2.42	1.64	6.00	30.70
2017	13.64	12.88	2.37	1.14	2.18	32.20

Source: House of Commons Library

The figures are given in millions.

These numbers have been rounded to two decimal places (the nearest 10 000 voters). Rounding can sometimes result in anomalies.

The trend for the total number of voters between 2010 and 2017 is **upwards** (or **increasing**).

You may be asked to comment on a **trend**, which is normally **upwards** or **downwards** but can also be **flat**.

Worked example

tier F

The table shows average house prices in January by UK country from 2011 to 2017.

	Average house price (£000s)						
Country	**2011**	**2012**	**2013**	**2014**	**2015**	**2016**	**2017**
England	174	174	177	188	203	220	232
Wales	128	127	128	132	136	140	146
Scotland	130	126	123	127	135	137	138
Northern Ireland	119	105	97	104	111	119	125

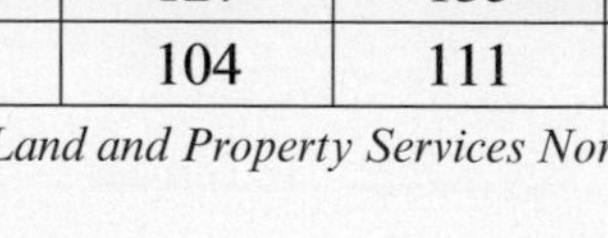

Source: HM Land Registry, Registers of Scotland, Land and Property Services Northern Ireland and Office for National Statistics

(a) Write down the average house price in Scotland in January 2014. **(1)**

£127 000

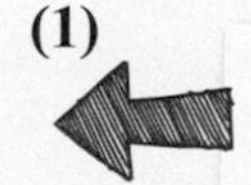

Read along the Scotland row and down the 2014 column.

(b) Describe the trend in the average house price in Wales from January 2011 to January 2017. **(2)**

The trend was flat from January 2011 to January 2013 and upwards from January 2013 to January 2017.

Small changes, such as the change from 2011 to 2012 to 2013, are too small to indicate a trend.

Now try this

The table shows the percentage of the population aged 65 or over for five countries from 2012 to 2016.

Country	2012	2013	2014	2015	2016
Germany	20.9	21.0	21.1	21.2	21.4
United Kingdom	16.8	17.2	17.5	17.8	18.0
Italy	21.2	21.6	22.0	22.4	22.7
Japan	24.3	25.0	25.7	26.3	26.9
United States	13.6	14.0	14.4	14.8	15.2

Source: dataworldbank.org

For the population aged 65 or over:

(a) Which country has the lowest percentage for the five years? **(1)**

(b) Which countries had a percentage over 20% in 2012? **(1)**

(c) Which country had the greatest increase in the percentage from 2012 to 2016? **(1)**

Two-way tables

A **two-way table** shows information about two categories of data.

Some values in a two-way table will be given; others may need to be calculated using addition or subtraction.

This two-way table shows people's replies when they were asked if they preferred sponge pudding with custard or with cream.

No one was allowed to pick both custard and cream in their reply.

Remember: data that has two variables is called **bivariate data**.

34 adults preferred custard.

	With custard	With cream	Total
Adult	34	18	52
Child	22	23	45
Total	56	41	97

The figures in blue show the replies.
The figures in red are worked out from the table.

Completing a two-way table

This two-way table shows the numbers of male and female musicians in each section of an orchestra.

To complete a two-way table look for rows or columns with only one missing value.

Top row total is 23 + 17 = 40

	Strings	Wind and brass	Total
Male	23	17	40
Female	34	8	42
Total	57	25	82

42 − 34 = 8 so '8' must go in female wind and brass.

Worked example

tier F&H

A group of 53 students were asked what their main source of news was – TV or the internet. There were 32 boys and 21 girls in the group. 31 students, of which 18 were boys, said TV.

(a) Complete the two-way table. **(2)**

	TV	Internet	Total
Boys	18	14	32
Girls	13	8	21
Total	31	22	53

(b) What was the students' main source of news? **(1)**

TV. The table shows that 9 more students used TV as their main source of news than used the internet.

In this case you have to fill in a two-way table from the information given. Make sure you use all the information given in the question.

Begin by putting in the values you are given. Don't forget the 53 in the first sentence.

Look for rows or columns with a single gap, and work out the missing numbers.

Check that the numbers in the rows and in the columns add up to the totals.

Always give your answer in context and with a reason.

Now try this

tier F&H

A teacher asked 30 students if they had a school lunch, a packed lunch or if they went home for lunch.
17 of the students were boys. 4 of the boys had a packed lunch. 7 girls had a school lunch.
3 of the 5 students who went home were boys.

(a) Record this information in a two-way table. Use the headings 'School', 'Packed' and 'Home'. **(3)**

(b) Work out the number of students who had a packed lunch. **(1)**

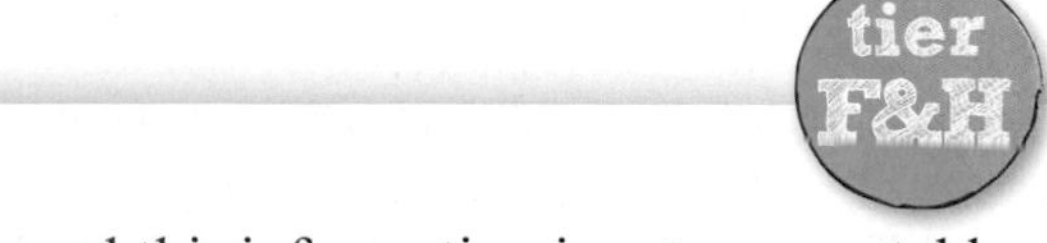

Had a look ☐ Nearly there ☐ Nailed it! ☐

Pictograms

Pictograms are a way of summarising data in a chart. You need to be able to draw and interpret pictograms.

This pictogram shows information about sales from a shop.

You will be expected to deal with halves and quarters when interpreting pictograms.

Pictograms may not be suitable for large numbers as symbols often cannot be easily divided.

Week 1	3 bottles
Week 2	1 bottle and 1 quarter bottle
Week 3	3 bottles and 1 half bottle
Week 4	1 bottle

Key: 1 bottle symbol represents 12 bottles

- $3 \times 12 = 36$ bottles sold in Week 1.
- This half symbol means $12 \div 2 = 6$ bottles. So 42 bottles were sold in Week 3.
- The **key** shows how many each symbol stands for.

Worked example

tier F

The pictogram gives some information about the numbers of cheeses Pippa sold in her shop in March, in April and in May.

March	3 circles
April	2 circles and 1 half circle
May	4 circles
June	1 circle and 1 half circle

Key: 1 circle represents 10 cheeses

Pippa sold 30 cheeses in March.

(a) Complete the key. **(1)**

Pippa sold 15 cheeses in June.

(b) Complete the pictogram. **(1)**

The price of a cheese is £18.

(c) Work out the total price of the cheeses Pippa sold in May and June. **(1)**

4 circles in May gives 40 cheeses.

$40 + 15 = 55$

Total price $= 55 \times £18 = £990$

Use the information given to work out the key. Pippa sold 30 cheeses in March and there are 3 shapes in the 'March' row, so 1 shape represents 10 cheeses.

A full shape represents 10 cheeses so a half shape represents 5 cheeses. Draw 1 full shape and 1 half shape to show 15 cheeses.

You need to be prepared to use the information given in a pictogram to carry out further calculations. Add together the numbers of cheeses sold in May and June then multiply by £18.

Now try this

tier F

1 The table shows the numbers of boxes of chocolates sold on four days of one week.

Monday	Tuesday	Wednesday	Thursday
4	12	6	10

Draw a pictogram for this information. Use the symbol ⊞ to represent four boxes. **(3)**

2 The pictogram gives information about the numbers of pets kept by residents in one street.

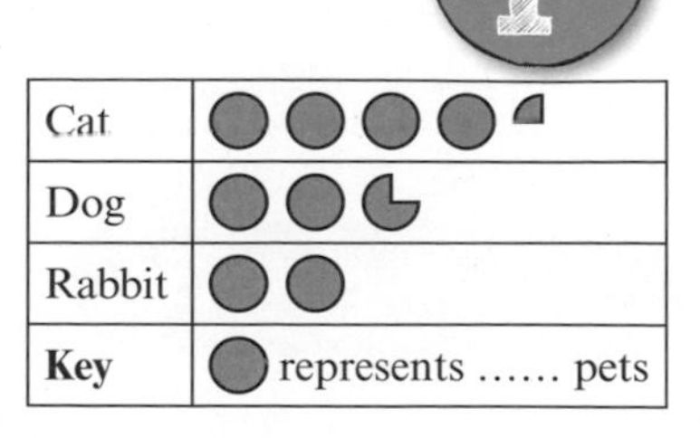

17 of the pets kept by the residents are cats.

(a) Complete the key for the pictogram. **(1)**

(b) Work out the number of dogs and the number of rabbits that are kept by the residents of the street. **(2)**

Bar charts 1

Bar charts and **vertical line graphs** are a good way of representing **discrete** data given in a tally chart or frequency table.

They can also be used to represent qualitative data. You met these types of data on page 1.

The table shows information about the types of trees in a wood.

Type of tree	Frequency
Oak	6
Ash	8
Pine	4
Beech	2

The bar chart shows how this information can be displayed.

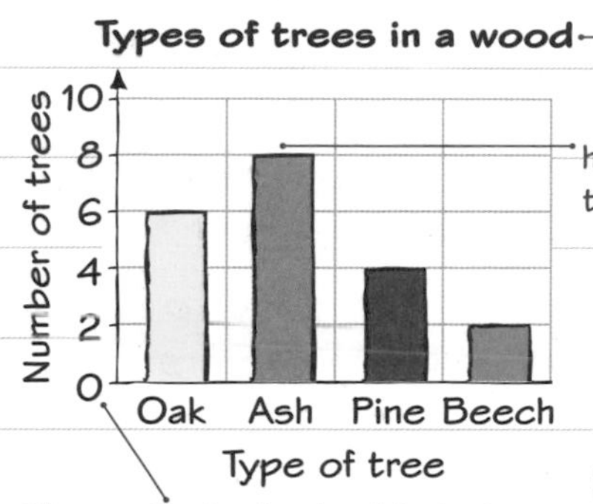

The bar chart should have a title.

In a bar chart the height (or length) of the bars represents the frequency.

The bars should all be the same width and equally spaced.

Both axes must be labelled. You can label the vertical axis 'Frequency'.

The vertical axis should start from 0 and go up by equal amounts each division.

Worked example

This vertical line graph (stick graph) gives information on the town people prefer to shop in.

(a) How many people said Bath? **(1)**

12

(b) More people preferred Bristol than Swindon. How many more? **(1)**

15 − 8 = 7

(c) How many people were asked altogether? **(1)**

12 + 15 + 10 + 8 = 45

The line that goes up to between 14 and 16 must be a whole number – so 15.

A line graph is similar to a bar chart but it has very thin bars.

Now try this

tier F&H

1 The table gives information about the numbers of vehicles on a road one morning.

Car	Van	Lorry	Bus	Other
8	7	11	3	5

Draw a bar chart to show this data. **(2)**

Make sure you start the scale at 0 and go up in equal steps.

2 The bar chart shows the sales of different types of tea in a tea shop in one day.

Sales of tea

Number sold: 0, 10, 20, 30, 40

Lemon, Green, Breakfast, Black, Herbal

(a) How many boxes of black tea were sold? **(1)**

(b) What is the difference between the smallest number of boxes and the greatest number of boxes of tea sold? **(2)**

Had a look ☐ Nearly there ☐ Nailed it! ☐

Bar charts 2

Multiple bar charts have more than one bar for each class.

Composite (or compound) bar charts group several different bars into a single bar and often show percentages.

Multiple bar charts

This multiple bar chart shows the sales of three makes of cars in four quarters of one year.

There are four sets of three bars to show how sales change over the year.

Using this bar chart it is easy to see that sales of Seat cars were high in the first two quarters but then fell.

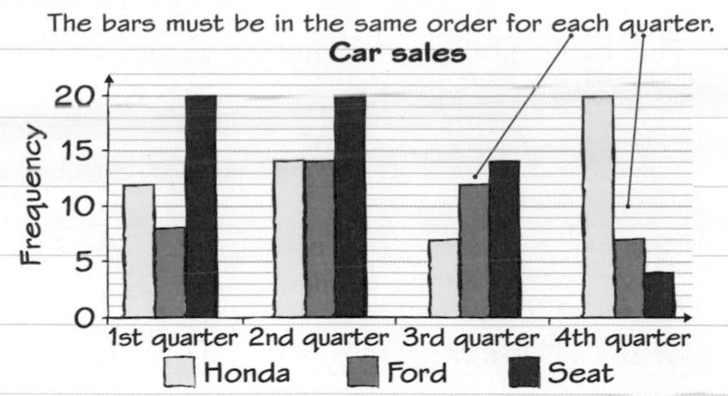

Composite bar charts

This composite bar chart shows how the percentages of men and women seen jogging have changed over two years.

The percentage of women has increased from 25% to 40%.

Composite bar charts can be harder to understand than multiple bar charts but do show the proportions within each group better.

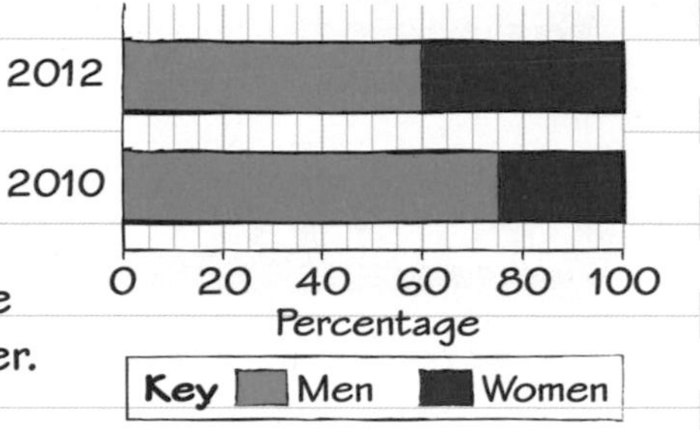

Worked example

tier F&H

The composite bar chart shows the percentages of families with 1, 2 and 3 children in a town.
Did the percentage of families with 2 children increase or decrease between 1980 and 2010? **(2)**

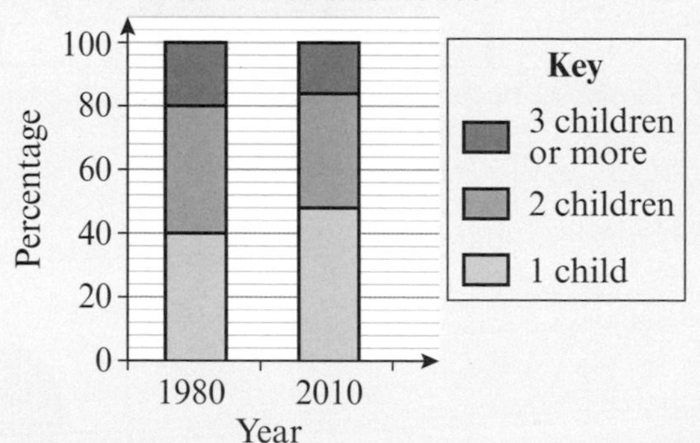

In 1980, the percentage was 80 − 40 = 40
In 2010, the percentage was 84 − 48 = 36
So the percentage decreased.

You need to be careful when you are reading off a composite bar chart. Use the key to work out which rectangle represents two children. If this rectangle doesn't start at 0 you will have to read off the values at the top and bottom of the rectangle and use subtraction.

Read off the percentage of families with 2 children in 1980 and in 2010. Write down both values, then write a conclusion saying whether the percentage increased or decreased.

Now try this

tier F&H

The table gives the times (in minutes) two boys spent watching TV on four days of one week.

Monday		Tuesday		Wednesday		Thursday	
Karim	Andy	Karim	Andy	Karim	Andy	Karim	Andy
40	30	30	50	85	70	140	170

(a) Draw a multiple bar chart to show this information. **(3)**
(b) Describe the trend in the times Andy spent watching TV. **(1)**

Don't forget to include a key.

Stem and leaf diagrams

The ordered display of individual discrete data values in a stem and leaf diagram shows the distribution of the data.

The diagram shows the numbers of emails 15 people received one day.

The **key** is necessary to interpret the diagram.

The **leaves** must always be single digits.

The column with 0, 1, 2 and 3 is the stem.

The rows contain the leaves.

In this case the tens make the stem and the units are the leaves.

0	6	8				
1	2	2	4	6		
2	0	2	3	4	7	7
3	1	3	5			

Key 3 | 5 represents 35 emails

Back-to-back stem and leaf diagrams

Back-to-back stem and leaf diagrams show two sets of data with the same stem. The smallest values on each row are next to the stem.

This diagram shows the marks of a group of students in two tests.

In a back-to-back stem and leaf diagram, the keys can be combined.

Test 1				Test 2		
6	3	1	5	7	9	9
	4	2	6	3	8	
8	0	0	7	5	5	
	6	3	8	7	7	9

Key: 6 | 5 = 56 5 | 7 = 57

6 | 5 | 7 represents 56 marks on the left and 57 marks on the right.

Worked example

tier F&H

Here are the numbers of goals a player has scored in her last 20 netball games.

23 9 20 14 23 6 17 24 24 18
16 10 22 21 11 8 21 15 8 22

Draw an ordered stem and leaf diagram to show this data. **(3)**

0	9	6	8	8				
1	4	7	8	6	0	1	5	
2	3	0	3	4	4	2	1	1	2

Use 0, 1 and 2 for the stem. Draw an unordered diagram first.
Cross off each number in the original list as you copy it onto the diagram.

0	6	8	8	9				
1	0	1	4	5	6	7	8	
2	0	1	1	2	2	3	3	4	4

Then copy the stem and rewrite the leaves on each row in order of increasing size.

Key 2 | 0 represents 20 goals

Remember to include the units in the key.

Now try this

tier F&H

1 Here are the heights, in cm, of 16 stalks of wheat. Show the heights on a stem and leaf diagram. **(2)**

120 131 108 110 122 132 127 105
122 133 121 137 112 119 104 113

Start the stem with 10, 11 and so on.
The leaves should always be single digits.

2 These are the weights, in kg, of 10 male babies and 10 female babies born on the same day in one hospital.

Male babies										Female babies									
5.0	4.3	4.5	3.9	4.6	4.8	3.7	3.5	3.8	3.6	2.7	3.2	4.1	2.9	3.3	3.8	3.2	3.3	3.0	3.6

(a) Show the weights on a back-to-back stem and leaf diagram. **(2)**

(b) Comment on the differences between the weights of the male and the female babies. **(1)**

Pie charts 1

Pie charts are generally used to show **qualitative** data. You need to be able to interpret them accurately. Remember that the angle of any sector in a pie chart is **proportional** to the number it represents.

Interpreting pie charts

This pie chart gives information about the replies that students gave to the question 'What is your most important subject?'

The pie chart shows that 'English' got the most votes and 'Maths' got one quarter of the votes.

Most important subject

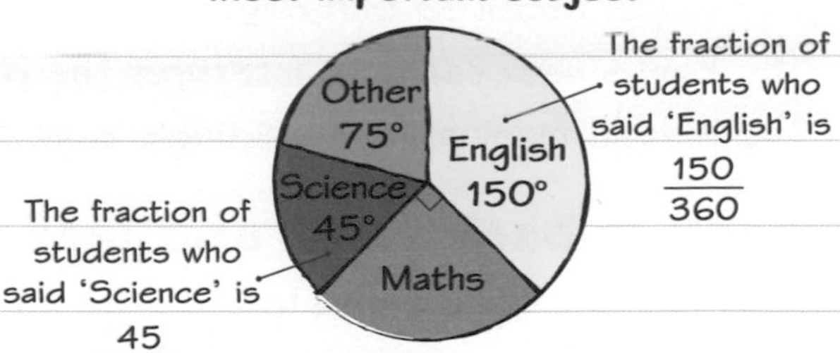

Using a formula

You can use this formula to work out what each sector represents:

$$\text{Number represented} = \frac{\text{angle of sector}}{360} \times \text{total}$$

LEARN IT!

Worked example

tier F&H

The pie chart gives information from a survey about where people worked. There were 135 people in the survey.

Place of employment

Other
Swindon
Bath 128°
West Wilts 96°
Bristol 64°

(a) Work out how many people worked in Bath. **(2)**

For Bath: $\frac{128}{360} \times 135 = 48$ people

There were 15 people in the Other category.

(b) Work out the size of the angle to represent Other on the pie chart. **(2)**

Angle for Other $= \frac{15}{135} \times 360 = 40°$

(c) Work out how many people worked in Swindon. **(2)**

Angle for Swindon $= 360° - 128° - 64° - 96° - 40° = 32°$

Number for Swindon $= \frac{32}{360} \times 135 = 12$ people

In the exam you might also need to measure angles on pie charts and use the formula.

Remember that the sum of the angles in a pie chart must be 360°.

Now try this

tier F&H

The pie chart gives information about the masses of vegetables grown on a farm. The total mass of vegetables produced was 2000 kg.

(a) Work out the mass of potatoes produced. **(2)**

(b) Work out the total mass of the beans and leeks. **(2)**

The mass of marrows produced was 800 kg.

(c) Work out the size of the angle to represent marrows on the pie chart. **(2)**

Production of vegetables

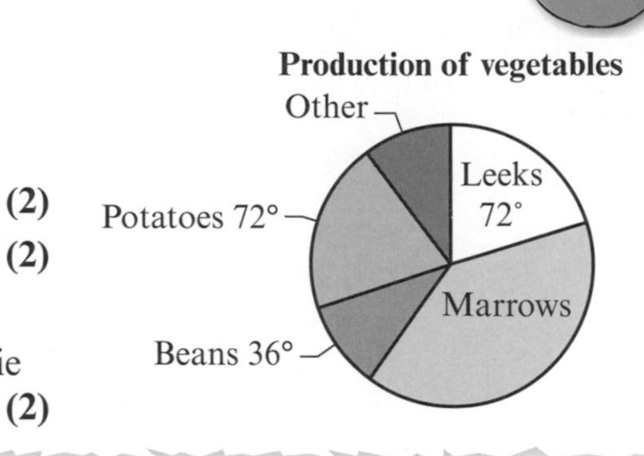

Pie charts 2

You need to be able to draw pie charts accurately from given data. You will need a pair of compasses, a protractor and a ruler.

Working out angles for pie charts

The sizes of the sectors in a pie chart are proportional to the numbers they represent.

Blood type	Number of people	Angle
O	30	$\frac{30}{72} \times 360 = 150°$
A	24	$\frac{24}{72} \times 360 = 120°$
B	12	$\frac{12}{72} \times 360 = 60°$
AB	6	$\frac{6}{72} \times 360 = 30°$
	72	

Use this rule to work out the angles for pie chart:

$$\text{Angle} = \frac{\text{number to be represented}}{\text{total}} \times 360°$$

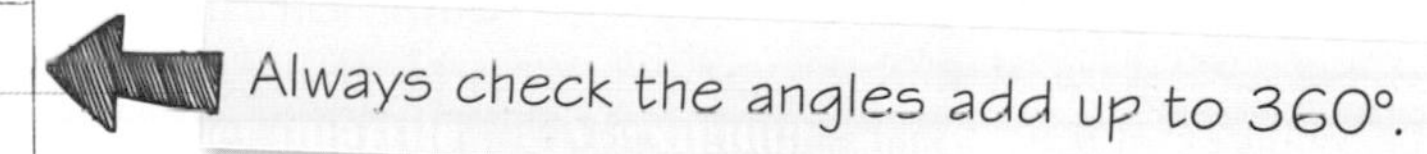
Always check the angles add up to 360°.

Add the numbers of people to find the total.

Worked example

tier F

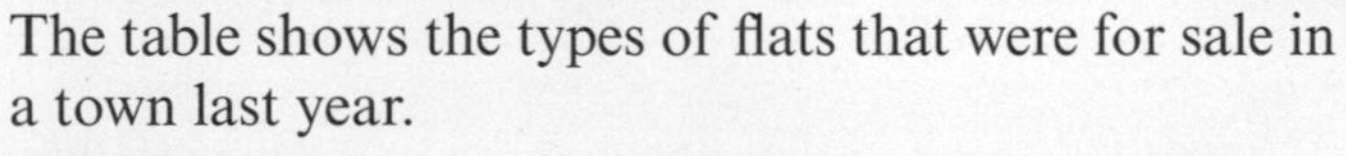

The table shows the types of flats that were for sale in a town last year.

Type	Bedsit	1-bed	2-bed	3-bed
Number	90	42	30	18

Draw a pie chart to show this information. **(3)**

Total = 90 + 42 + 30 + 18 = 180 flats

So the angle for bedsits $= \frac{90}{180} \times 360° = 180°$

The other angles are 84°, 60°, 36°

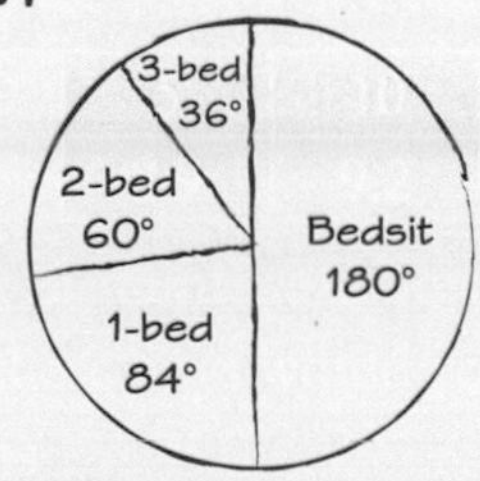

Work out the total first.

The angle for 'Bedsits' $= \frac{90}{\text{total}} \times 360°$

Work out the other angles.

Use a pair of compasses, a protractor and a ruler to draw the sectors.

Don't forget to **label** the sectors.

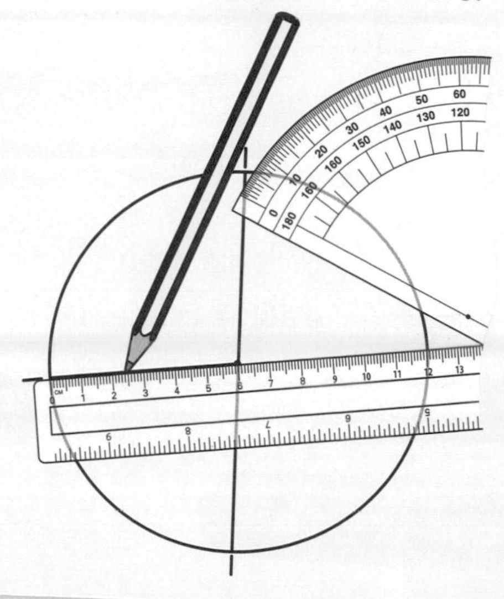

Now try this

tier F

The table shows the types of dairy cows on a large farm.

Type of cow	Holstein	Friesian	Jersey	Ayrshire
Number	80	40	36	24

Draw a pie chart to show this information. **(3)**

Comparative pie charts

Comparative pie charts are used to **compare** the numbers in populations of **different sizes.**

These two pie charts show preferred sources for news for men and women.

The pie chart for women is larger than the pie chart for men because the total number of women sampled is greater.

Men

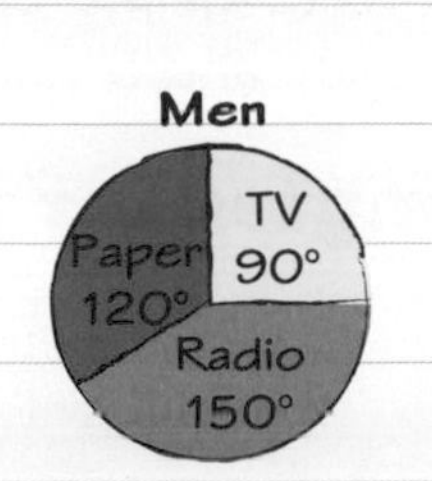

Women

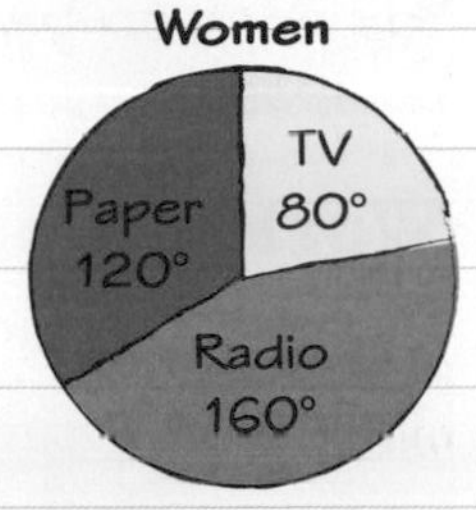

$$\frac{\text{area of large pie chart}}{\text{area of small pie chart}} = \frac{\text{total number of women}}{\text{total number of men}}$$

Making a comparison

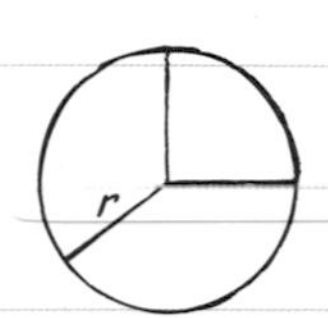

Total number $= n$

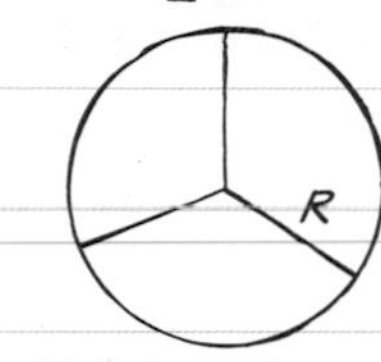

Total number $= N$

$$\frac{\text{area of large pie chart}}{\text{area of small pie chart}} = \frac{\pi R^2}{\pi r^2} = \frac{R^2}{r^2} = \frac{N}{n}$$

LEARN IT!

So, use $\frac{R^2}{r^2} = \frac{N}{n}$ where r is the radius of the small circle, R is the radius of the large circle, n is the total number in the small sample and N is the total number in the large sample.

Worked example

tier H

The pie charts show the holiday destinations of a sample of 200 people in 2016 and a sample of people in 2017.

Work out how many more people went to France in 2017 than in 2016. **(3)**

Holiday destinations 2016: Other 90°, UK 120°, Spain 78°, France 72°. Radius = 3 cm

Holiday destinations 2017: Other 138°, UK 90°, Spain 60°, France 72°. Radius = 4.5 cm

The total number of people in 2017 = N

$\frac{R^2}{r^2} = \frac{N}{n}$ so $\frac{R^2}{r^2} = \frac{4.5^2}{3^2} = \frac{N}{200}$

$N = \frac{20.25}{9} \times 200 = 450$

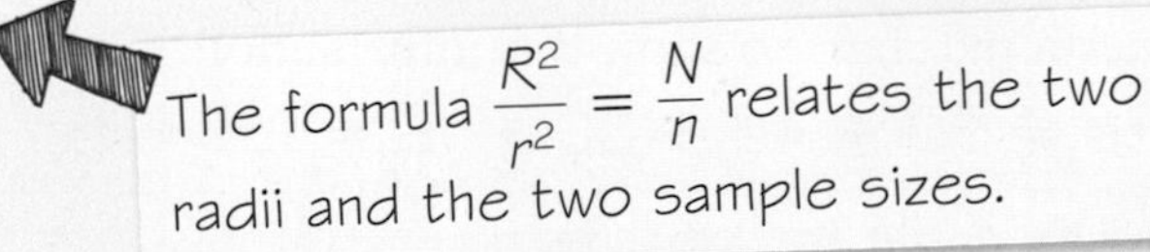

Number who went to France in 2016

$= \frac{72}{360} \times 200 = 40$

Number $= \frac{\text{angle}}{360} \times$ total

Number who went to France in 2017

$= \frac{72}{360} \times 450 = 90$

50 more people went to France in 2017 than in 2016.

Although the angle for France is the same in both pie charts, the number it represents is different because the pie charts are **different sizes.**

Now try this

tier H

Leo is studying the distribution of weeds in marshland and in moorland. The size of the marshland sample is 200. The size of the moorland sample is 450.

Leo draws a pie chart of radius 4 cm for the marshland.

(a) Calculate the radius of the comparative pie chart for the moorland. **(2)**

Leo works out that the angle for one type of weed in the pie chart for the marshland sample is the same as the angle for the same type of weed in the moorland sample.

(b) Describe how the number of this type of weed in the marshland sample compares with the number of the same type of weed in the moorland sample.
Explain how you reach your conclusion. **(2)**

Population pyramids

Population pyramids are similar to bar charts or stem and leaf diagrams.
They give information about the age structure of a population.
The bar charts are presented horizontally with male and female populations on opposite sides.

The diagram shows an estimation of the population (in thousands) in Edinburgh in mid-2016.

In the 21–40 age group there were just under 90 000 males and just over 90 000 females.

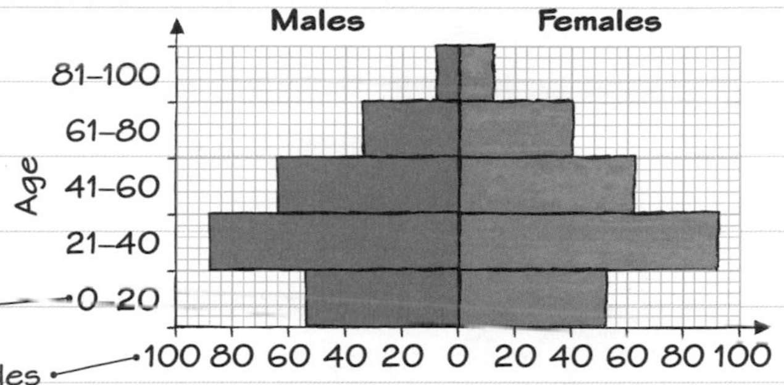

The vertical scale always has the youngest ages at the bottom.

The scales on both sides must be the same.

Source: Office for National Statistics.

Worked example

tier F&H

The population pyramid shows the estimated distribution of males and females in the UK in mid-2016 as a percentage of the total number of each gender.

(a) What percentage of males were aged under 30? **(1)**

18.5 + 20 = 38.5%

(b) What percentage of females were aged 45 or over? **(1)**

20.5 + 15.5 + 9.5 = 45.5%

Read off the percentages for females aged 45–59, 60–74 and 75 and over and add them together.

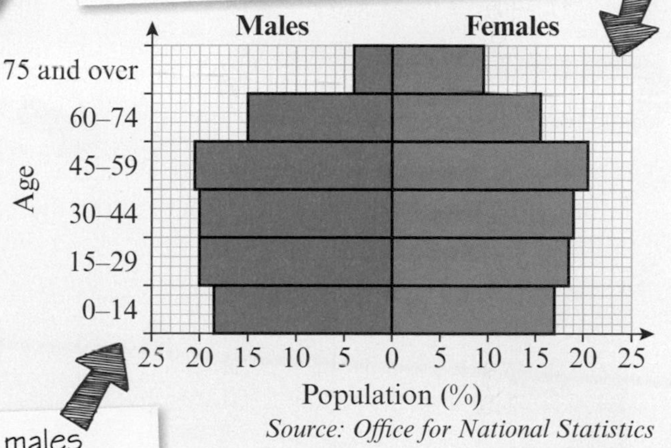

Source: Office for National Statistics

Read off the percentages for males aged 0–14 and aged 15–29 and add them together.

Now try this

tier F&H

Look at the population pyramid in the worked example above and this population pyramid, which shows the predicted distribution of the UK population in 2050.

Compare the estimated population in 2016 with the predicted population in 2050. **(2)**

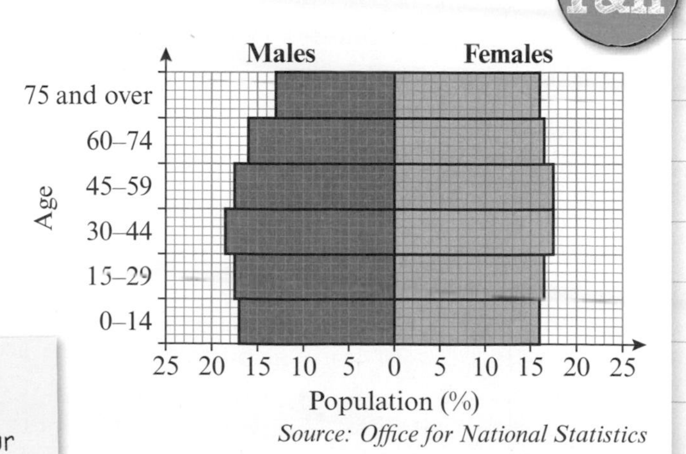

Source: Office for National Statistics

Make two different comments comparing the percentages of the populations in different age groups. Give actual percentages to back up your comments. You can use **skew** to describe the distributions – there is more about this on page 56.

Had a look ☐ Nearly there ☐ Nailed it! ☐

Choropleth maps

A **choropleth map** uses different colours or shading to show how data varies across different **geographic** areas. This choropleth map shows the distribution in the UK of the surname 'Davies'.

You need to be able to interpret and compare choropleth maps.

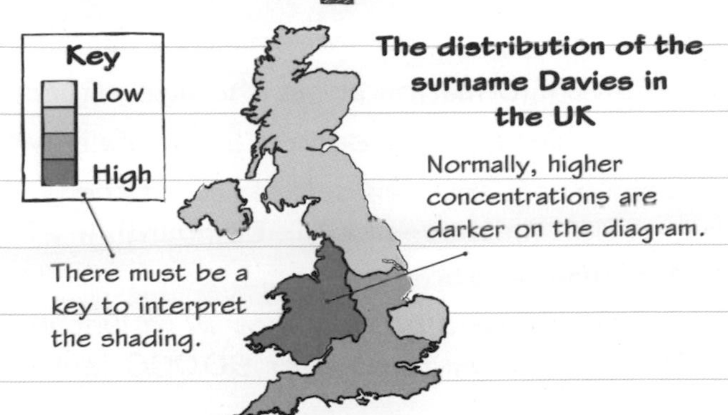

Choropleth maps on grids

You can use a choropleth map to display the change in population over an area.

This diagram shows the masses of worms that live in the topsoil of an 8 m by 12 m plot.

In your exam, choropleth maps will use different shades of grey or shading patterns to show different values.

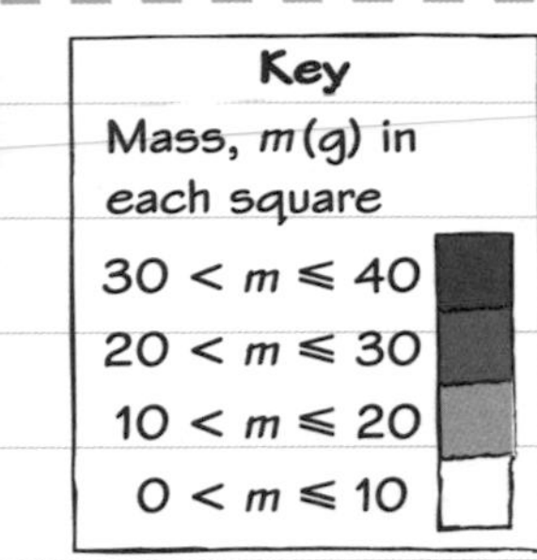

Worked example

tier F

You will be given a key showing how to shade for different values.

The diagram shows a field split into 2-metre squares. The numbers in each square show the numbers of flowers.

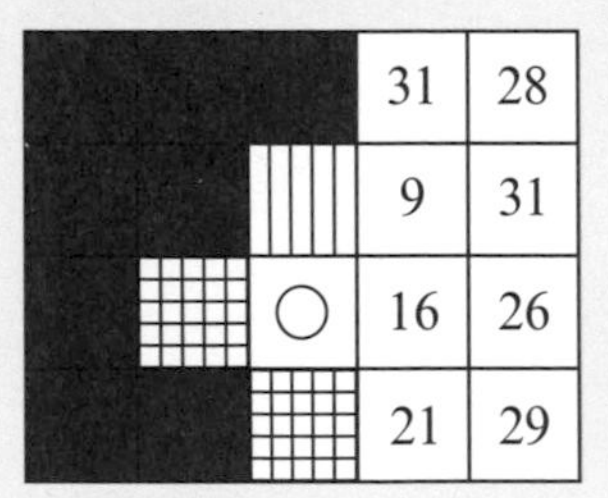

Key: Number of flowers

31 – 40
21 – 30
11 – 20
0 – 10

You cannot use colour on an exam paper, so different degrees of shading from white to black are used, with black being the most dense area.

(a) On the grid provided complete the choropleth. **(2)**

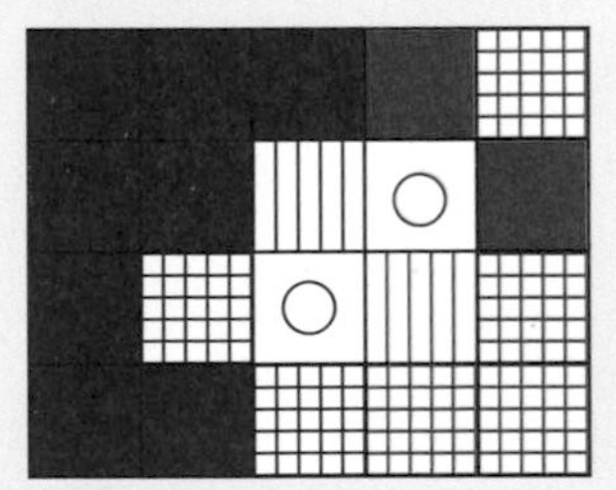

(b) Describe the distribution of the flowers. **(1)**

There is a higher density of flowers on the edges of the field.

You can make any general comment which describes the spread.

Now try this

tier F

The diagram shows a square field. The numbers in the small squares show the numbers of insects.

140	139	311	312
126	320	211	189
330	326	201	116
300	270	170	80

See what the highest and lowest values are before deciding on a key. Then think of suitable intervals – it's usual to have four.

(a) Draw a choropleth map to show this data. You must include a suitable key. **(2)**

(b) Describe the distribution of the insects. **(1)**

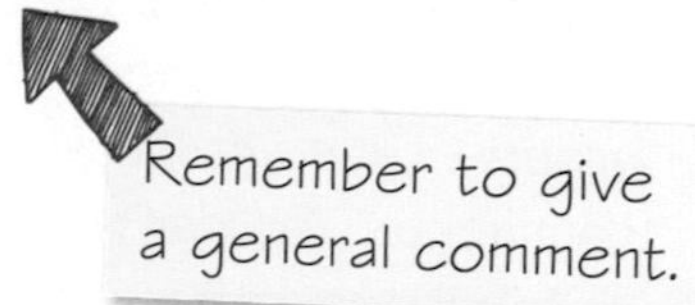

Histograms and frequency polygons

You use a **histogram** to represent **grouped continuous data**. There are **no gaps** between the bars on a histogram (unless one of the intervals has a zero frequency).

A **frequency polygon** is formed by joining the midpoints of the tops of the bars in a histogram. It is a useful way to show the shape of a distribution.

Interpreting histograms

You can read frequencies from the histogram as well as work out cumulative frequencies.

This histogram shows the distances people could throw a rock.

It shows that there were 5 people who threw between 4 m and 5 m.

You can work out that there were 5 + 8 = 13 people who threw up to 6 m.

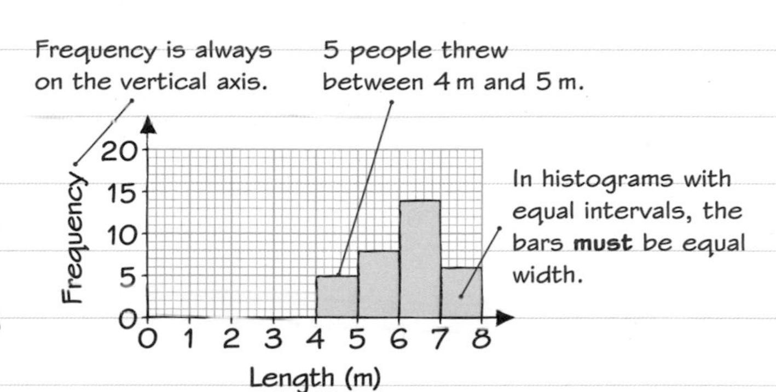

Worked example

tier F&H

The table gives some information about the heights of some children in a club.

Height, h (cm)	Frequency
$110 < h \leqslant 120$	8
$120 < h \leqslant 130$	13
$130 < h \leqslant 140$	16
$140 < h \leqslant 150$	10
$150 < h \leqslant 160$	7

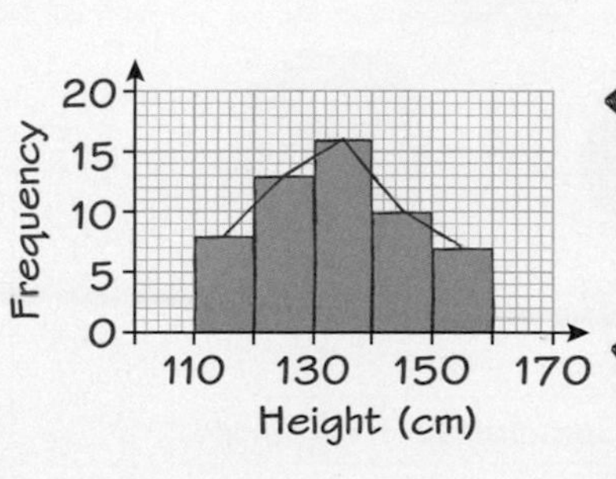

(a) Draw a histogram for this data. **(2)**

(b) Draw a frequency polygon for this data. **(1)**

Draw each bar to the given frequency. Use a ruler and a sharp pencil whenever you are drawing graphs in the exam.

Draw the frequency polygon by joining the **midpoints** of the tops of the bars with straight lines.

Now try this

tier F&H

1 The table shows information about the widths of a sample of flowers in a bunch of flowers.

Width, W (cm)	$0 < W \leqslant 2$	$2 < W \leqslant 4$	$4 < W \leqslant 6$	$6 < W \leqslant 8$	$8 < W \leqslant 10$
Frequency	12	9	7	4	2

(a) Draw a histogram to show the information in the table. **(2)**

(b) Draw a frequency polygon to show the information in the table. **(1)**

Remember there must be no gaps between the bars.

2 The histogram gives information about the heights of plants in a garden.

How many plants had a height of 40 cm or more? **(2)**

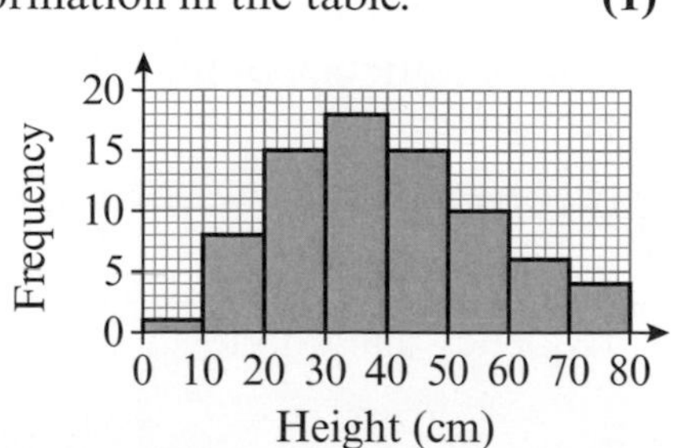

Had a look ☐ Nearly there ☐ Nailed it! ☐

Cumulative frequency diagrams 1

Cumulative frequency is a running total of frequencies in a frequency table.

If the data is continuous, the points plotted on the **cumulative frequency diagram** or **graph** can be joined with straight lines or a smooth curve.

Cumulative frequency step polygons

To represent discrete data which can only take certain values (often whole numbers) you use a cumulative frequency step polygon.

The table shows the numbers of cars in a car park at noon on 38 days.

Number of cars	11	12	13	14	15
Frequency	8	11	9	6	4

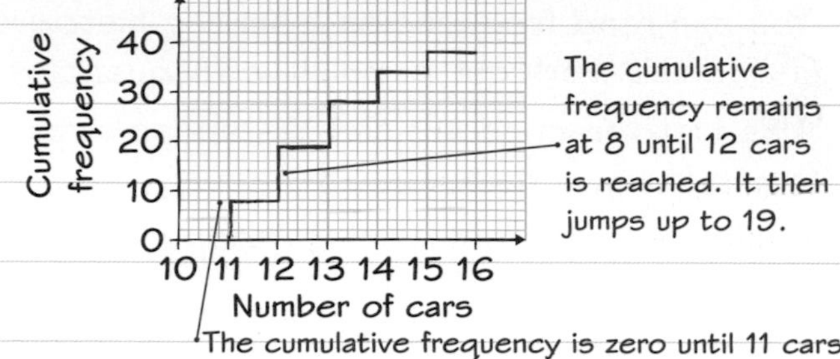

This is the **cumulative** table for the same data.

Number of cars	≤11	≤12	≤13	≤14	≤15
Cumulative frequency	8	8 + 11 = 19	19 + 9 = 28	28 + 6 = 34	34 + 4 = 38

There were up to 13 cars on 28 out of the 38 days.

Worked example

tier F&H

The table gives information about the heights of 100 students.

Height, h (cm)	Frequency	Cumulative frequency
$120 < h \leqslant 130$	8	8
$130 < h \leqslant 140$	16	8 + 16 = 24
$140 < h \leqslant 150$	24	24 + 24 = 48
$150 < h \leqslant 160$	32	48 + 32 = 80
$160 < h \leqslant 170$	20	80 + 20 = 100

Draw a cumulative frequency diagram to show this information. **(3)**

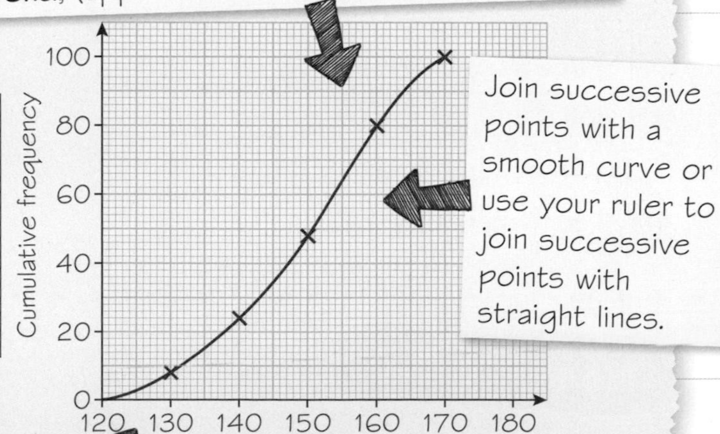

Now try this

tier F&H

1 The table shows some information about the snowfall each day in January and February one year.

Snowfall, S (cm)	$0 < S \leqslant 2$	$2 < S \leqslant 4$	$4 < S \leqslant 6$	$6 < S \leqslant 8$	$8 < S \leqslant 10$
Frequency	20	14	12	8	6

Draw a cumulative frequency diagram to show this information. **(3)**

Start by working out the cumulative frequencies.

Cumulative frequency diagrams 2

You need to be able to interpret **cumulative frequency diagrams** or **graphs**. The diagrams can be used to **estimate** values.

The worked example shows how to use the plotted information in a cumulative frequency diagram to make estimations.

Worked example

tier F&H

The diagram gives some information about the costs of some laptops.

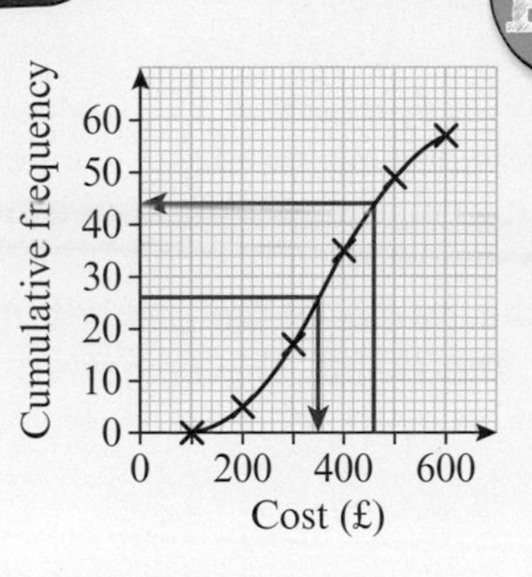

(a) Find an estimate for the number of laptops with a cost of more than £460. **(1)**

57 − 44 = 13

Draw a vertical line from 460 on the horizontal axis up to the curve.
Read off the value on the cumulative frequency axis (44).
Remember to subtract from the total number (57).

(b) Find an estimate for the maximum cost of the cheapest 26 laptops. **(1)**

£350

Draw a horizontal line from 26 across to the graph.
Read off the value from the cost axis.

Now try this

tier F&H

1 The diagram gives the cumulative frequency percentages of ages of people in India.

Cumulative frequency (%)
100
80
60
40
20
0
0 20 40 60 80 100
Age (years)

(a) Find an estimate for the percentage of people who are aged 20 or less. **(1)**

(b) What age is exceeded by 25% of the people in India? **(1)**

Remember to use a ruler and sharp pencil when you draw lines for reading off.

2 Sylvie bought some packets of sweets. The table shows the total number of sweets in each packet.

Number of sweets	28	29	30	31	32	33
Frequency	4	7	8	12	16	3

(a) Draw a cumulative frequency step polygon to display this data. **(3)**

(b) How many packets of sweets did Sylvie buy? **(1)**

(c) What percentage of the packets contained more than 30 sweets? **(1)**

Had a look ☐ Nearly there ☐ Nailed it! ☐

The shape of a distribution

You need to be able to identify the shape of distributions of data. The **shape of a distribution** is the shape formed by, for example, a stem and leaf diagram, a frequency polygon or the bars of a histogram. A distribution can be **symmetrical**, have **positive skew** or have **negative skew**.

The diagrams show an example of a histogram for each of the three different cases. They illustrate how the shape of the distribution determines the skew.

Positive skew

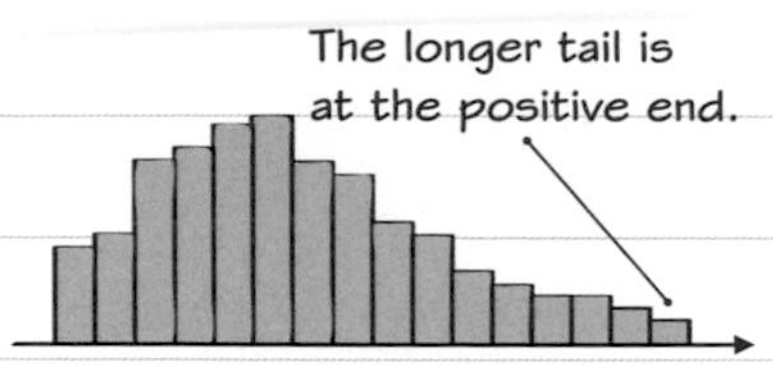

Most of the data values are at the lower end and the distribution is stretched out in the positive direction →.

Symmetrical

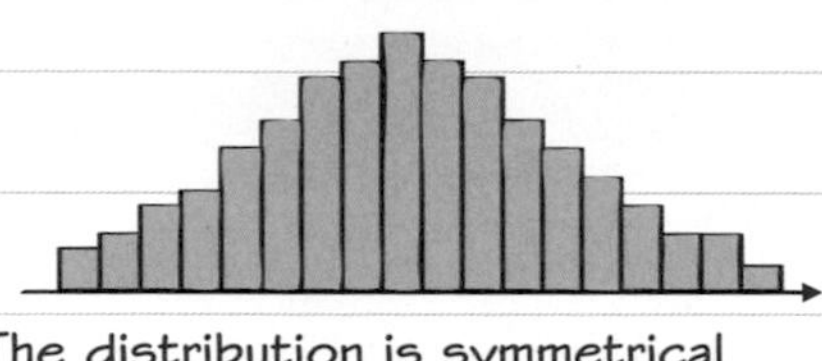

The distribution is symmetrical about the middle and has no skew.

Negative skew

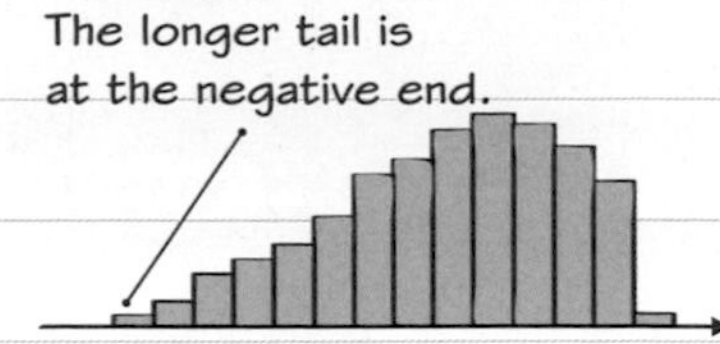

Most of the data values are at the upper end and the distribution is stretched out in the negative direction ←.

A distribution has two **tails**. The tails are the parts of the distribution that are furthest away from the mean. The positive tail is on the right, the negative tail is on the left.

Worked example

tier F&H

The data shows the pulse rates of 10 people before and after exercise.

Pulse rate before	69 90	68 82	74 81	76 95	75 87
Pulse rate after	88 102	89 104	97 86	95 111	92 96

(a) Draw a back-to-back stem and leaf diagram to display the data. **(3)**

Pulse rate before			Stem	Pulse rate after			
	9	8	6				
6	5	4	7				
7	2	1	8	6	8	9	
	5	0	9	2	5	6	7
			10	2	4		
			11	1			

Key: 6 | 7 = 76 8 | 9 = 89

(b) Comment on the shape of each distribution. **(2)**

Pulse rate before exercise is approximately symmetrical.
Pulse rate after exercise has positive skew.

(c) Draw grouped frequency polygons on the same axes to compare the pulse rate before with the pulse rate after. **(3)**

Plot the frequency for the pulse rate, r, at the middle of the grouped intervals:
$60 \leqslant r < 70$
$70 \leqslant r < 80$ etc.

Frequency
after exercise
before exercise
0 1 2 3 4
60 70 80 90 100 110
Pulse rate (bpm)

Now try this

tier F&H

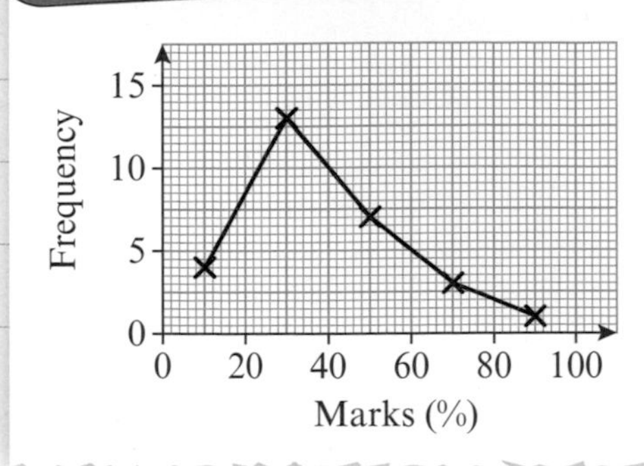

The frequency polygon shows the marks of a group of students in a test.

Comment on the shape of the distribution. **(1)**

Histograms with unequal class widths 1

Histograms with **unequal class widths** use **frequency density** on the vertical axis.

This histogram shows the times taken by some students to solve a problem. The **area** of each bar represents a **number** of students.

The first bar shows that 8 (1 × 8) students took less than 1 minute. The fourth bar shows that 7 (0.5 × 14) students took between 3.5 and 4 minutes.

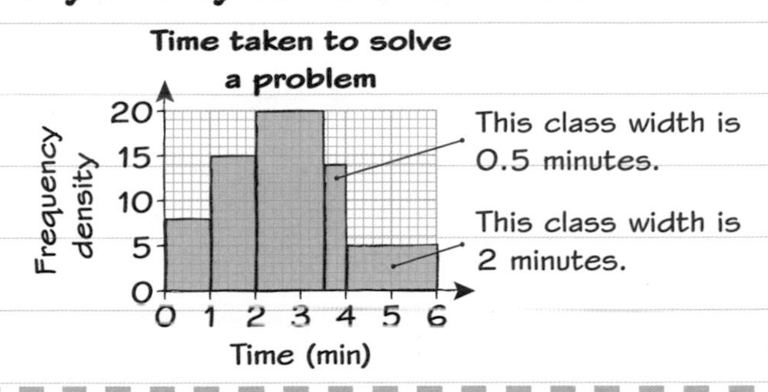

Frequency density

Before drawing a histogram you need to calculate the **frequency density** for each class interval.

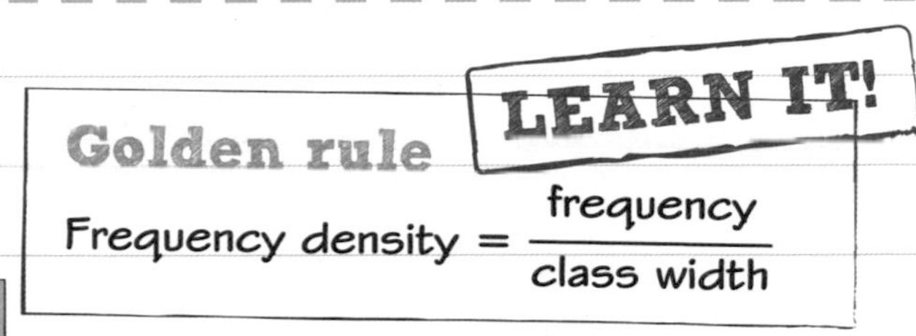

The table shows the lengths of leaves in a garden.

Length, L (cm)	Frequency	Class width	Frequency density
$0 < L \leq 5$	12	5	$12 \div 5 = 2.4$
$5 < L \leq 15$	10	10	$10 \div 10 = 1.0$
$15 < L \leq 25$	16	10	$16 \div 10 = 1.6$
$25 < L \leq 45$	12	20	$12 \div 20 = 0.6$
$45 < L \leq 60$	6	15	$6 \div 15 = 0.4$

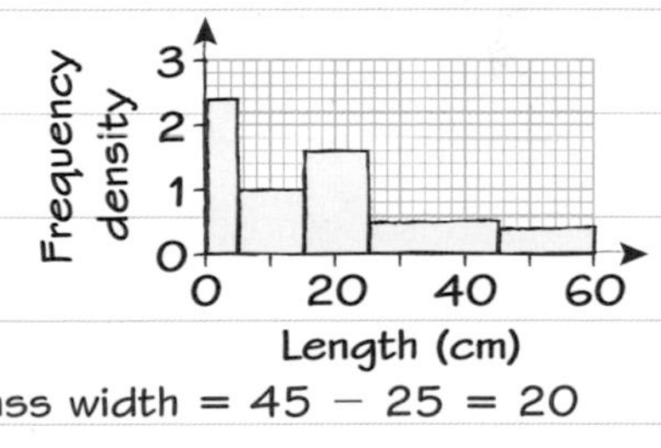

Class width = 45 − 25 = 20

Worked example

tier H

The table gives data about the times (in minutes) some trains were late.

Time, t (min)	Frequency	Class width	Frequency density
$0 < t \leq 5$	12	5	$12 \div 5 = 2.4$
$5 < t \leq 10$	18	5	$18 \div 5 = 3.6$
$10 < t \leq 20$	14	10	$14 \div 10 = 1.4$
$20 < t \leq 40$	12	20	$12 \div 20 = 0.6$

Draw a histogram to show this data. **(3)**

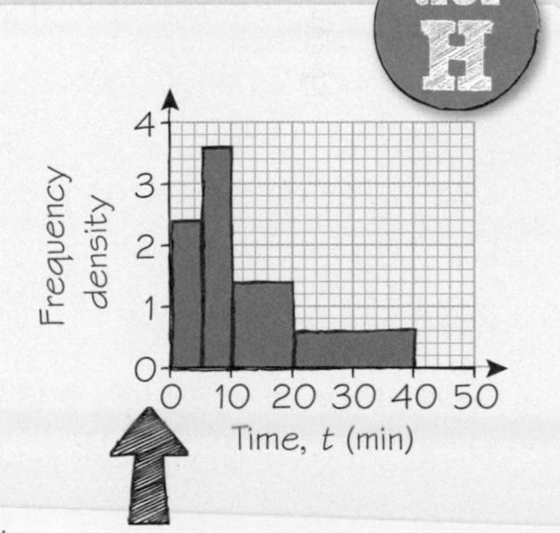

1. Work out each class width.
2. Use the golden rule to work out each frequency density.

3. Label the vertical axis Frequency density.
4. Decide on the scale for each axis (these may be given to you in the exam).
5. Draw the bars.

Now try this

tier H

The table gives information about the calorie intake per day of some people.

Number (N) of calories	$1000 < N \leq 1750$	$1750 < N \leq 2000$	$2000 < N \leq 2250$	$2250 < N \leq 2750$	$2750 < N \leq 4000$
Frequency	16	34	42	54	14

Draw a histogram to show this information. **(3)**

Histograms with unequal class widths 2

You can find frequencies from histograms with unequal class widths by using frequency density or by using an area key.

Histograms and area keys

Histograms can be drawn or interpreted by using an **area key**.

In this histogram, one small square represents $50 \div 25 = 2$ fish.

The number represented by each bar in the histogram can be found by using the area key.

For example, the number of fish with length 15 cm or less is given by the area of the first two bars.

Area $= 10 \times 5 + 16 \times 2.5 = 90$ small squares

Number of fish $= 90 \times 2 = 180$

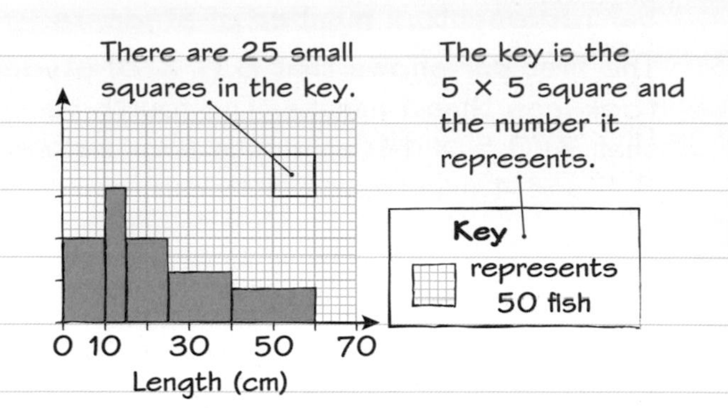

Worked example

tier H

The histogram shows the times (in seconds) people spent at a self-service checkout.

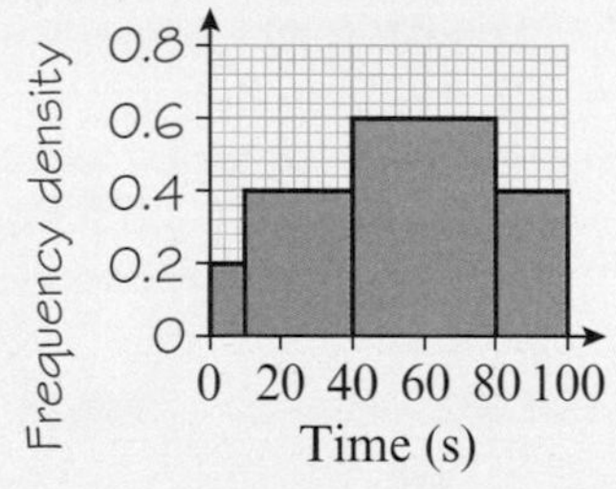

Time, t (s)	Frequency
$0 < t \leqslant 10$	2
$10 < t \leqslant 40$	12
$40 < t \leqslant 80$	24
$80 < t \leqslant 100$	8

Another method is to use the information you have been given to create an area key.

▦ represents 4 people

Use the histogram to complete the frequency table marks. **(3)**

The frequency density of the $80 < t \leqslant 100$ class interval is $\frac{8}{20} = 0.4$

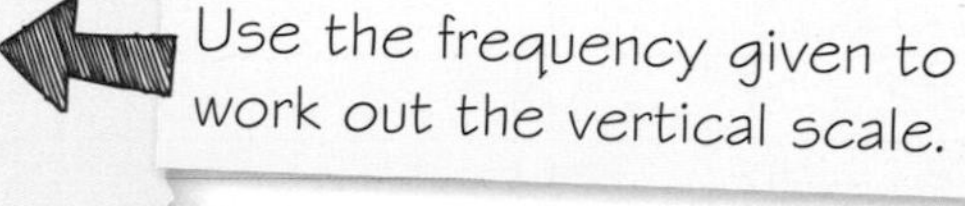

Frequency of the $0 < t \leqslant 10$ class $= 0.2 \times 10 = 2$

Frequency of the $10 < t \leqslant 40$ class $= 0.4 \times 30 = 12$

Frequency of the $40 < t \leqslant 80$ class $= 0.6 \times 40 = 24$

Frequency = frequency density × class width

Now try this

The histogram gives information about the water intake per day of some plants.

Water intake, W (ml)	$100 < W \leqslant 150$	$150 < W \leqslant 200$	$200 < W \leqslant 300$	$300 < W \leqslant 500$
Frequency	200	300	450	540

(a) Draw a histogram to show this information. Use a horizontal scale of 1 cm = 100 ml. Use an area key of 1 cm^2 represents 200 plants. **(4)**

(b) Estimate the percentage of plants which had a water intake of more than 250 ml per day. **(3)**

Misleading diagrams

You need to be able to recognise when diagrams are **misleading**.

Sometimes graphs or diagrams are deliberately misleading, but sometimes the way data is presented can be unintentionally misleading.

Comparisons in three-dimensional diagrams can be difficult.

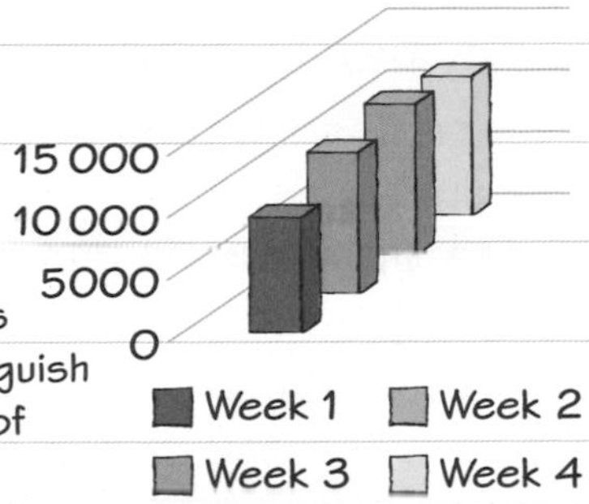

15 000
10 000
5000
0
Week 1
Week 2
Week 3
Week 4

In a 3D bar chart, it is very difficult to distinguish between the heights of the bars.

Things to watch out for:

- vertical scales that are too big or too small, are not linear or do not start at zero
- axes that are not labelled clearly
- missing features, such as a key
- 3D diagrams, which may distort the features.

In a 3D pie chart the angles are distorted and sections at the back may appear smaller than those at the front.

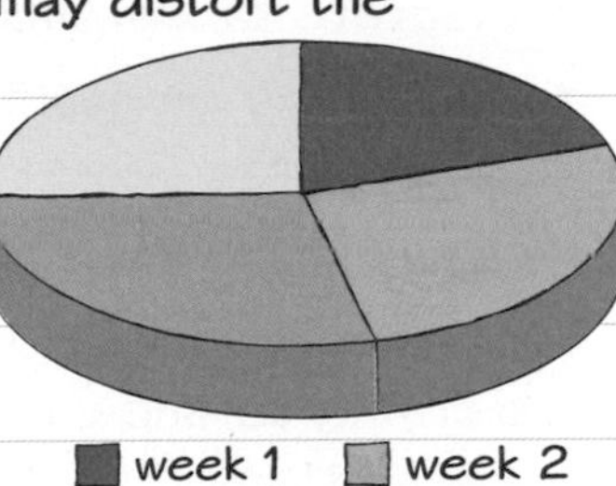

week 1
week 2
week 3
week 4

In pictograms, images of different sizes may be used, which makes it impossible to compare values.

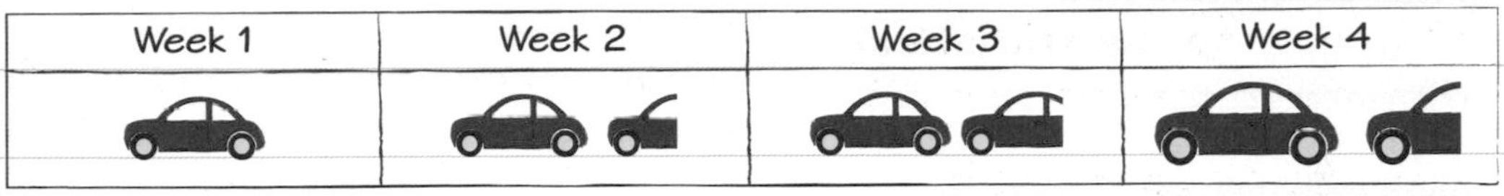

Week 1
Week 2
Week 3
Week 4

Worked example

tier F&H

Write down **two** reasons why this graph may be misleading.

Scales that do not start at zero and use a large scale give a misleading impression about the difference between the plotted points.

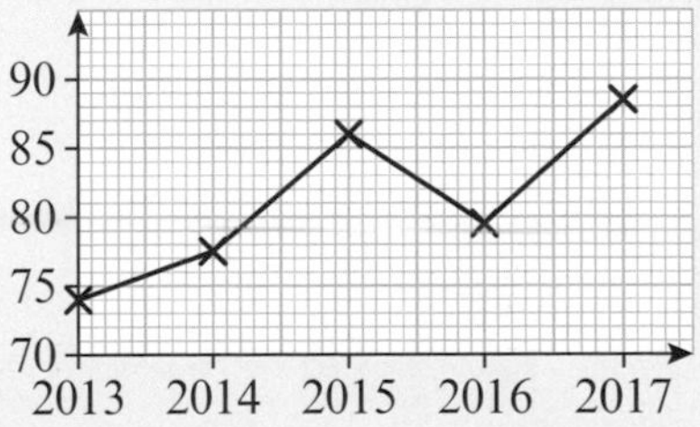

90
85
80
75
70
2013
2014
2015
2016
2017

Axes without labels prevent you from knowing what the data represents.

(2)

The vertical scale does not start at zero.
The axes are not labelled.

Now try this

tier F&H

1 Write down **three** reasons why this graph may be misleading.

350
250
100
0
Males
Females

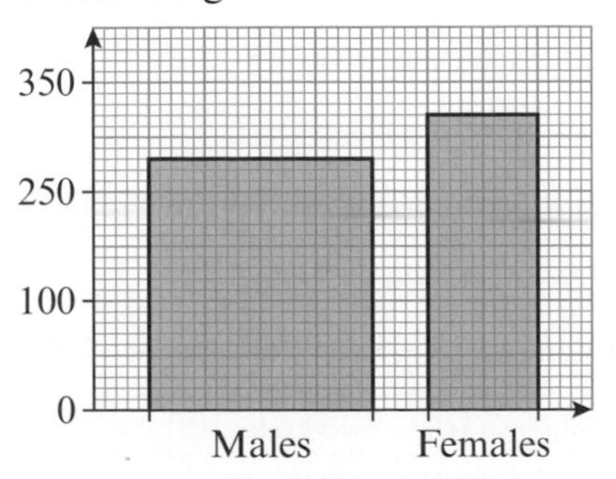

(3)

2 This histogram was drawn to show the marks of students in an exam.

15
10
5
0
Frequency
50
55
65
75
80
Marks

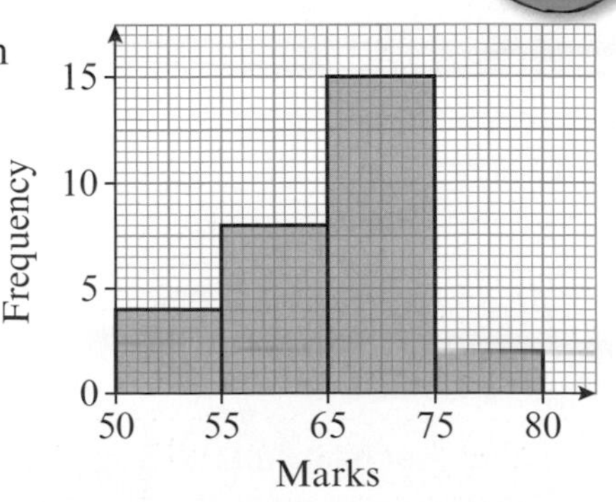

(a) **Write down three** things that are wrong in the diagram. **(3)**

(b) **Draw** an accurate histogram for the data. **(2)**

Had a look ☐ Nearly there ☐ Nailed it! ☐

Choosing the right format

You need to be able to interpret and compare data sets presented in different formats as well as **choose an appropriate format** to represent data and explain your choice.

Presenting data

You need to think about the **type of data** you want to present. Is it

- discrete or continuous?
- qualitative or quantitative?
- grouped?

You also need to consider your **target audience** and what you are trying to show in the presentation.

You should know when it is appropriate to use **statistical software** to present your data.

Common ways of displaying data

A **table** shows raw data presented in rows and columns. It can be used to display discrete, continuous, qualitative or quantitative data. A table gives exact values but it can be hard to identify trends and patterns.

A **bar chart** can be used to display discrete, continuous, qualitative or quantitative data. It can be used to compare data and show trends and patterns. Bar charts can be drawn in single, multiple or composite form.

A **histogram** is used to display continuous grouped data. The area of the bars represents the frequency. It can also be used to compare data and show trends and patterns.

Pie charts are used to display proportion but not totals. You can use **comparative pie charts** to compare data with different totals using radii which are in proportion.

Worked example

tier F&H

Karla completed a survey on the average waiting time it took patients to get a non-urgent appointment with a doctor at a local practice. Write a comment on the three different ways of presenting her results. **(3)**

Number of weeks (w)	Number of patients
$w < 1$	215
$1 \leqslant w < 2$	275
$2 \leqslant w < 3$	220
$3 \leqslant w < 4$	100
$4 \leqslant w < 5$	12
$w \geqslant 5$	8

A

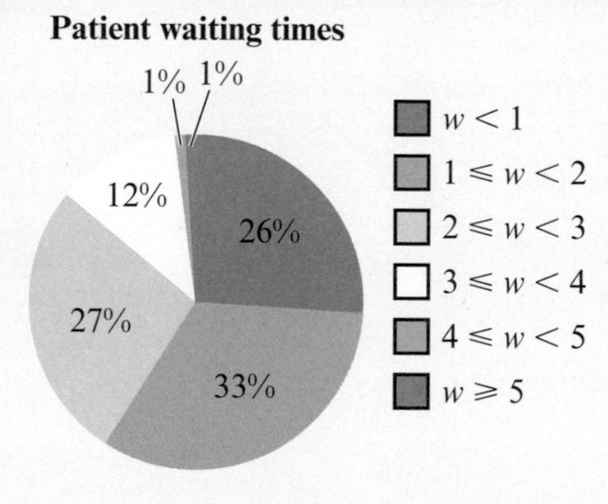

B

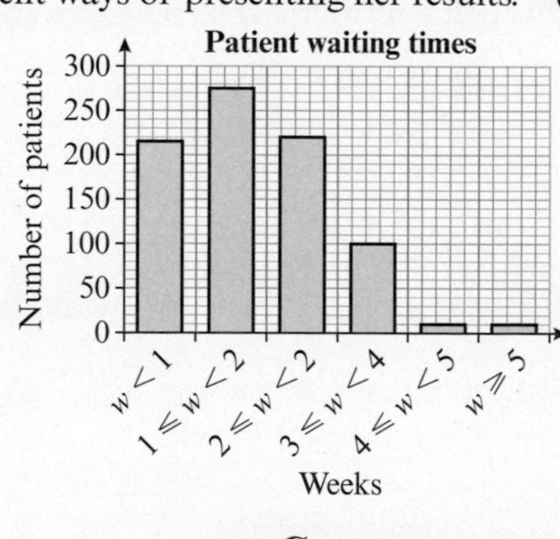

C

A The table shows the number of patients in each category clearly and it is easy to see the category with the highest number of patient responses.

B The pie chart shows the proportion of patients in each category but you cannot identify the number of patients in each case.

C The bar chart shows the trend of the responses but the scale makes it difficult to identify the number of patients in the last two categories.

Now try this

tier F&H

These are the results of a survey on the number of mobile phone messages students sent in one day.

Number of texts	0–10	11–30	31–60	61–100	101–200
Number of students	9	35	58	49	22

(a) Explain why a histogram is not a suitable format for this data. **(1)**

(b) Represent this data in a suitable format. Explain your choice. **(3)**

Averages

An **average** is a single value used to describe a set of data as a measure of **central tendency**.

The **mode**, **median** and **mean** are the three averages you need to know for your exam.

- The **mode** is the value appearing **most** often.
- The **median** is the **middle** value when all the values are in order.
- To work out the **mean**, add up all the values and divide by the number values.

The mode is the height of any of these people.

The median is the height of this person in the middle.

In order of increasing height

Order the data from smallest to largest. Make sure you have the same number of values in the ordered list as in the original list.

You can use the rule $\frac{n+1}{2}$ to find where middle value of n values is.

There are 9 data values. $\frac{9+1}{2} = 5$, so the 5th value is the median.

Worked example (tier F)

Here is a list of the numbers of children in nine families.

6 3 2 2 3 1 1 8 1

(a) Find the mode. **(1)**

The mode is the most common so the mode is 1.

(b) Find the median. **(1)**

In order: 1 1 1 2 2 3 3 6 8

The median is the middle value when written in order so the median is 2.

Median

You need to be careful when you are calculating the median of an **even** number of data values.

You can use the rule that the median is the $\frac{n+1}{2}$th value.

Here is a list of the number of cars in a small car park over a period of 14 days.

14 16 18 15 14 18 13 20 22 19 12 12 21 25

In order:

~~12~~ ~~12~~ ~~13~~ ~~14~~ ~~14~~ ~~15~~ (16 18) ~~18~~ ~~19~~ ~~20~~ ~~21~~ ~~22~~ ~~25~~

$\frac{n+1}{2} = \frac{14+1}{2} = 7.5$ The 7.5th value is halfway between the 7th and 8th values.

The 7th number is 16 and the 8th number is 18 so the median is 17 (halfway between 16 and 18).

Arithmetic mean

The **arithmetic mean**, usually called simply the **mean**, is the sum of all the values divided by the number of values.

For a set of n data values $x_1, x_2, \ldots, x_n$

Mean $= \bar{x} = \frac{\sum x}{n}$ where

- $\bar{x}$ is the mean of all the x-values
- $\sum x$ is the sum of all the x-values.

This list shows the numbers of pupils in a primary school each morning for eight days.

152 165 165 163 159 160 160 158

$\sum x = 152 + 165 + 165 + 163 + 159 + 160 + 160 + 158 = 1282$

Mean $= \bar{x} = \frac{1282}{8} = 160.25$ There are 8 values in the list.

Now try this (tier F)

1 Here is a list of the numbers of patients a doctor saw in the first 10 working days of a new job.

18 16 26 18 15 17 23 18 24 28

(a) Find (i) the mode, (ii) the median, (iii) the mean. **(4)**

On the 11th day, the doctor saw 20 patients.

(b) Will the mean for all 11 days be greater or less than the mean for the first 10 days? Give a reason. **(1)**

Had a look ☐ Nearly there ☐ Nailed it! ☐

Averages from frequency tables 1

You need to be able to find the **mode** and **median** from a frequency table for **discrete** data.

Mode

The mode is the value with the **highest frequency** (the value that appears most often). Be careful – the mode is the **data value** and **not** the frequency.

The table gives information about the numbers of cars parked in the grounds of some buildings.

Number of cars	0	1	2	3	4	5
Frequency	5	16	12	10	7	3

The mode is the number of cars with the greatest frequency – so the mode is 1.

Median

The median is the middle value when the data is written in order of size.

In a frequency table the data is already in order.

Make an extra column headed Cumulative frequency.

Fill in the column by starting with 5.

The median is found by using the rule

Median = $\frac{n+1}{2}$ th data value

where n is the total frequency.

Number of cars	Frequency	Cumulative frequency
0	5	5
1	16	21
2	12	33
3	10	43
4	7	50
5	3	53

Add a cumulative frequency column.

Write down the first frequency.

5 + 16 = 21 goes here.

The final number in this column should equal the total frequency.

The median is the $\frac{53+1}{2}$ th value.

The 27th value is 2, so median is 2.

Worked example

tier F&H

The table gives information about the numbers of children in some families.

Number of children	Frequency	Cumulative frequency
0	8	0 + 8 = 8
1	11	8 + 11 = 19
2	10	19 + 10 = 29
3	7	29 + 7 = 36
4	4	36 + 4 = 40

Add a cumulative frequency column to the table to help with part **(b)**.

(a) Write down the mode. **(1)**

The mode is 1.

The number of children with the highest frequency is 1.

(b) Work out the median. **(1)**

$\frac{n+1}{2}$ = 20.5. The median is 2.

Find the mean of the 20th and 21st values.

Now try this

tier F&H

The table shows class attendance.

Attendance	22	23	24	25	26
Frequency	8	12	15	12	12

(a) Write down the mode. **(1)**

(b) Work out the median. **(2)**

Averages from frequency tables 2

You need to be able to find the **mean** from a frequency table.

For **discrete data** in a frequency table, mean $= \bar{x} = \frac{\sum fx}{\sum f}$ **LEARN IT!**

- $\bar{x}$ is the mean of all the x-values
- $\sum f$ is the sum of all the frequency (f) values
- $\sum fx$ is the sum of all the $f \times x$-values.

Mean of discrete data from a frequency table

Use the formula: Mean $= \frac{\sum fx}{\sum f}$

Add another column to the right of the table for $f \times x$.

The table here gives information about European shoe sizes of 20 people.
The mean = 598 ÷ 20 = 29.9

Shoe size	Frequency	$f \times x$
28	8	8 × 28 = 224
30	7	7 × 30 = 210
32	3	3 × 32 = 96
34	2	2 × 34 = 68
	20	598

28, 30, 32 and 34 are the x-values.

These two columns are the original table.

This is $\sum f$ (the sum of the frequencies).

This is $\sum fx$.

Worked example

tier F

The table shows information about the numbers of letters in the first names of a group of 50 people.

Number of letters	Frequency	$f \times x$
3	2	2 × 3 = 6
4	5	5 × 4 = 20
5	14	14 × 5 = 70
6	19	19 × 6 = 114
7	10	10 × 7 = 70

(a) Work out the mean number of letters in the names of the 50 people. **(2)**

$$\frac{6 + 20 + 70 + 114 + 70}{50} = \frac{280}{50} = 5.6$$

(b) Another person called Simon joins the group. Will the mean of the letters in the first names in the group increase? You must give a reason for your answer. **(1)**

The mean will not increase because 'Simon' has 5 letters which is less than the old mean of 5.6

Start by adding a column for $f \times x$. Work out each $f \times x$ and write down the answer.
If $\sum f$ is not given in the question then work it out by adding the frequencies.
Use mean $= \frac{\sum fx}{\sum f}$ from the formulae sheet.

There is no need to do any calculations. Compare the mean with the number of letters in Simon's name.

A common mistake is to write 5 × 0 = 5 instead of 0.
Check your working carefully.

Now try this

tier F

The table gives information about the numbers of mobiles in some households.

Number of mobiles	0	1	2	3	4
Frequency	5	12	23	30	14

Calculate the mean. Give your answer correct to 1 decimal place. **(2)**

Had a look ☐ Nearly there ☐ Nailed it! ☐

Averages from grouped data 1

Exact values for the mode and median cannot be found for **grouped data** because you do not know the actual data values, only the class intervals they lie in.

Modal class and class containing the median

The **modal class** is the interval which has the highest frequency.

The table gives information about the lengths of time some trains were late.

The **class that contains the median** is found by using cumulative frequency.

The median is the $\frac{n+1}{2}$th data value in the table where n is the total of the frequencies.

The modal class is $0 < T \leqslant 5$ because it has the highest frequency.

Time late, T (minutes)	Frequency	Cumulative frequency
$0 < T \leqslant 5$	12	12
$5 < T \leqslant 10$	8	20
$10 < T \leqslant 15$	11	31
$15 < T \leqslant 20$	8	39
$20 < T \leqslant 25$	4	43

$\frac{n+1}{2} = \frac{43+1}{2} = 22$

The 22nd data value lies in the interval $10 < T \leqslant 15$

The lowest 20 times go up to here. The 21st and 22nd times are in the next interval.

Worked example

tier F&H

The table gives information about the ages of people going on an outing.

Age, N (years)	Frequency	Cumulative frequency
$5 \leqslant N < 15$	18	18
$15 \leqslant N < 25$	20	38
$25 \leqslant N < 35$	16	54
$35 \leqslant N < 45$	21	75
$45 \leqslant N < 55$	16	91

Calculate an estimate for the median.
Give your answer correct to 3 significant figures. **(3)**

Estimated median $= L + \left(\frac{\frac{1}{2}n + F}{f}\right)w$

Where

L = lower boundary of median class interval (25)

n = total number of values (91)

F = cumulative frequency of intervals before median class interval (38)

f = frequency of median class interval (16)

w = width of median class interval (10)

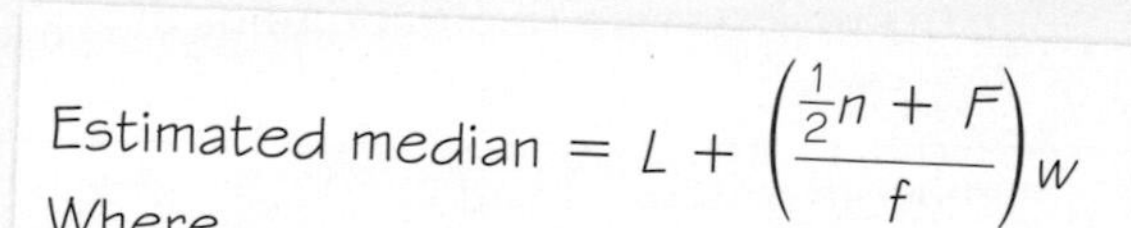

Median is the $\frac{91+1}{2} = 46$th data value.

Median class interval is $25 \leqslant N < 35$

Estimated median $= 25 + \left(\frac{\frac{1}{2}(91) - 38}{16}\right)10 = 29.7$ (3 s.f.)

Now try this

1 The table gives information about the numbers of days some people had off work.

Days off, D	$0 \leqslant D < 5$	$5 \leqslant D < 10$	$10 \leqslant D < 15$	$15 \leqslant D < 20$	$20 \leqslant D < 25$
Frequency	12	8	11	8	5

(a) (i) Write down the modal class. **(1)** (ii) Write down the class interval that contains the median. **(2)**

(b) Calculate an estimate for the median. Give your answer correct to 3 significant figures. **(3)**

Averages from grouped data 2

You can **estimate** the **mean** from a grouped frequency table by using the formula

Mean = $\frac{\sum fx}{\sum f}$ (See page 41 for more about the mean.) x represents the **midpoints of the intervals.**

You can use $\bar{x}$ ('x bar') as the symbol for the mean.

Mean of continuous data from a frequency table

The table gives information about the times some students spent on homework.

The midpoint of each interval is found by adding the end points and dividing by 2.

The midpoint of the interval $20 < T \leq 30$ is $x = \frac{20 + 30}{2} = 25$

Time, T (mins)	Frequency	Midpoint (x)	$f \times x$
$0 < T \leq 10$	12	5	$12 \times 5 = 60$
$10 < T \leq 20$	8	15	$8 \times 15 = 120$
$20 < T \leq 30$	3	25	$3 \times 25 = 75$
$30 < T \leq 40$	2	35	$2 \times 35 = 70$
	25		325

This is $\sum f$ (the sum of the frequencies). This is $\sum fx$.

Because you are using the midpoint of each interval, you are working out an **estimate** for the mean. You would need to know the time taken by every student to find the exact value.

Using the formula, $\bar{x} = \frac{\sum fx}{\sum f} = 325 \div 25 = 13$

Worked example

tier F&H

The table shows information about the numbers of years a group of people had been driving.
Calculate an estimate for the mean. **(3)**

Number of years of driving (N)	Frequency	Midpoint (x)	$f \times x$
$0 < N \leq 10$	8	5	$8 \times 5 = 40$
$10 < N \leq 20$	12	15	$12 \times 15 = 180$
$20 < N \leq 30$	14	25	$14 \times 25 = 350$
$30 < N \leq 40$	11	35	$11 \times 35 = 385$
$40 < N \leq 50$	5	45	$5 \times 45 = 225$
$\sum f = 50$			$\sum fx = 1180$

$\bar{x} = 1180 \div 50 = 23.6$ years

Work out the midpoints, x, of each interval.

Work out $f \times x$ for each interval and add to find $\sum fx$

Add up the frequencies to find $\sum f$

Use the formula $\frac{\sum fx}{\sum f}$ to find the mean.

You need to work out the midpoint of each interval first.

The table gives information about the areas, A (in m²) of plots of wasteland in a town.

Area, A (m²)	$0 < A \leq 200$	$200 < A \leq 400$	$400 < A \leq 600$	$600 < A \leq 800$	$800 < A \leq 1000$
Frequency	56	40	28	20	8

(a) Calculate an estimate for the mean area of a plot of wasteland. **(3)**
(b) Explain why your answer is an estimate. **(1)**

Had a look ☐ Nearly there ☐ Nailed it! ☐

Averages from grouped data 3

To calculate an **estimate** of the **mean** from a **histogram**, first set up a grouped frequency table and use the method on page 43. To calculate an estimate of the **median** either first set up a grouped frequency table or use **linear interpolation** on the histogram itself.

Estimating the median using a histogram

The histogram shows the numbers of hours students spent on homework in one week.

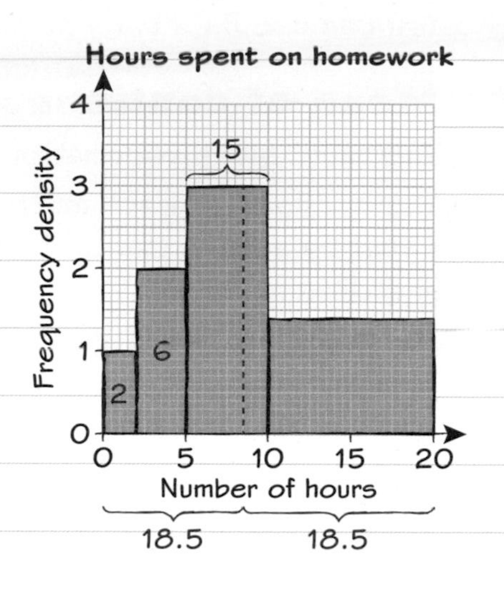

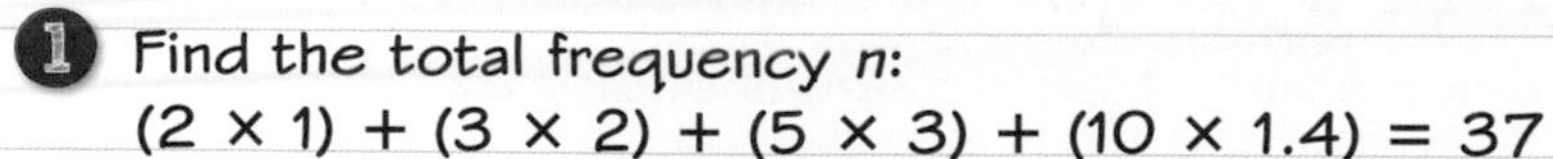

1. Find the total frequency n:
   $(2 \times 1) + (3 \times 2) + (5 \times 3) + (10 \times 1.4) = 37$
2. Find the position of the median, $\frac{1}{2}n$ and the bar this lies in:
   $\frac{37}{2} = 18.5$, in the class interval $5 < t \leqslant 10$

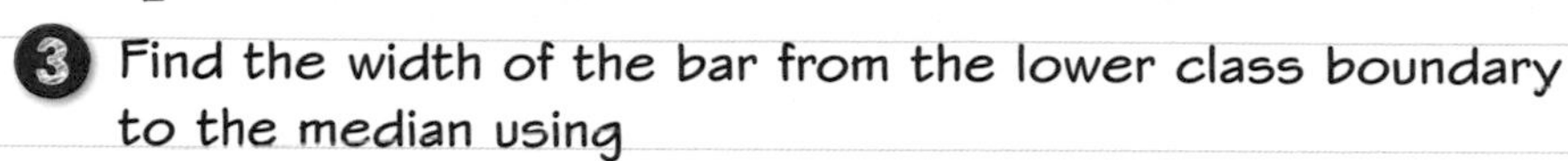

3. Find the width of the bar from the lower class boundary to the median using
   $$\text{Frequency density (fd)} = \frac{\text{frequency}}{\text{width}}$$
   Area of bar up to median value $= 18.5 - 2 - 6 = 10.5$
   $$3 = \frac{10.5}{\text{width}} \text{ so width} = \frac{10.5}{3} = 3.5$$

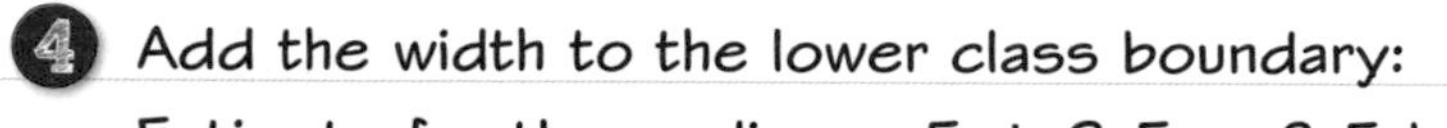

4. Add the width to the lower class boundary:

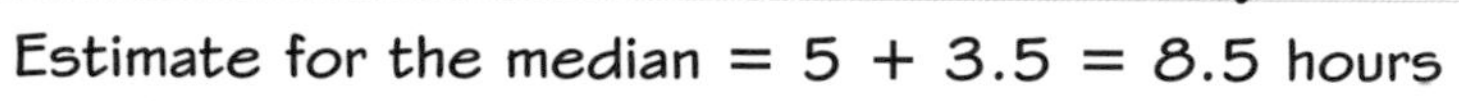

   Estimate for the median $= 5 + 3.5 = 8.5$ hours

Worked example

tier H

The histogram shows the ages of employees in a company.
Work out an estimate for the median age. **(2)**

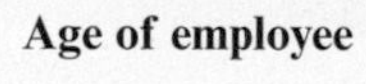

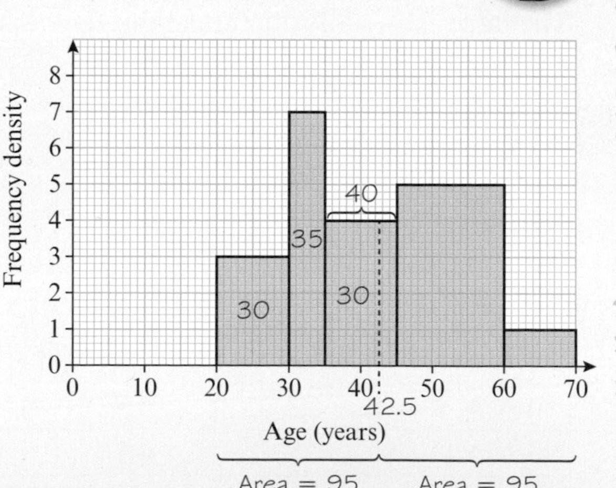

Total frequency $= 10 \times 3 + 5 \times 7 + 10 \times 4 + 15 \times 5 + 10 \times 1 = 190$

$190 \div 2 = 95$, in the class interval $35 < y \leqslant 45$

In $35 < y \leqslant 45$, area up to median value $= 95 - 35 - 30 = 30$

$\text{fd} = \frac{\text{frequency}}{\text{class width}}$, so $4 = \frac{30}{\text{class width}}$

class width $= \frac{30}{4} = 7.5$

Estimate for median $= 35 + 7.5 = 42.5$ years

Now try this

tier H

The histogram shows the heights of children in a drama group.
Work out an estimate of

(a) the mean height **(2)**

(b) the median height. **(2)**

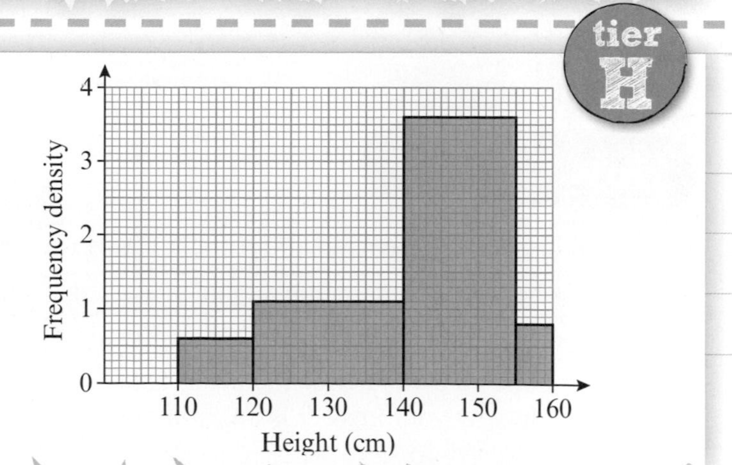

Transforming data

When all data values are **transformed** (increased or decreased) by the same amount or percentage, the averages are transformed by the same amount or percentage.

Calculating with transformed values

You can sometimes calculate the mean more easily if the data is transformed first.

This is a list of door heights:

2.05 2.02 2.14 2.01 2.20 2.09

To find the mean, these numbers can be transformed.

First subtract 2 from each value:

0.05 0.02 0.14 0.01 0.20 0.09

Then multiply by 100:

5 2 14 1 20 9

Mean of transformed numbers

$$= \frac{5 + 2 + 14 + 1 + 20 + 9}{6} = \frac{51}{6} = 8.5$$

Now reverse what you did to the numbers:

- divide by 100
- add 2.

Mean of original numbers $= \frac{8.5}{100} + 2 = 2.085$

Worked example

tier F&H

(a) Find the mean, median and mode for this list of prices. **(3)**

£45 £28 £36 £57 £28

Mean = (45 + 28 + 36 + 57 + 28) ÷ 5
= £38.80

In order: 28 28 (36) 45 57

Median = £36

Mode = £28

(b) The prices are increased by 20%. Find the new mean, median and mode for this data. **(2)**

New mean = £38.80 × 1.2 = £46.56

New median = £36 × 1.2 = £43.20

New mode = £28 × 1.2 = £33.60

To increase a value by 20% use the multiplier 1.2 to find 120%.

Worked example

These are the speeds, in seconds, recorded for a 400 m race.

59 52 56.5 59.5 58

Find the mean speed. **(2)**

Transformed data:

−50: 9 2 6.5 9.5 8

×10: 90 20 65 95 80

Mean of transformed data

$$= \frac{90 + 20 + 65 + 95 + 80}{5} = \frac{350}{5} = 70$$

Mean of original data $= \frac{70}{10} + 50 = 57$

Don't forget to apply the reverse of the transformation to the answer.

Now try this

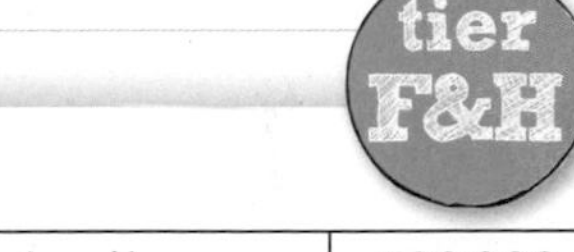

The table shows the numbers of visitors to the National Gallery each month in one year, correct to the nearest 1000.

Month	Visitors
April	582 000
May	562 000
June	433 000
July	560 000
August	459 000
September	309 000
October	436 000
November	499 000
December	446 000
January	434 000
February	564 000
March	579 000

Source: www.gov.uk

(a) Calculate the mean number of visitors per month. **(2)**

In the following year the visitor numbers increased by about 5%.

(b) What was the mean number of visitors per month in the following year? **(2)**

Using a transformation can save you time here, even though the question doesn't tell you to do so.

Geometric mean

The mean average of a set of n data values is sometimes called the **arithmetic mean** and involves the **addition** of n values. You also need to be able to calculate the **geometric mean** for a set of data. The geometric mean for n data values is the nth root of the **product** of the n values. It is a type of average which is commonly used for **growth rates** such as population growth or interest rates.

Calculating the geometric mean

LEARN IT!

Geometric mean = $\sqrt[n]{\text{value}_1 \times \text{value}_2 \times \ldots \times \text{value}_n}$

To calculate the geometric mean of 5, 7 and 12, there are 3 values, so $n = 3$

Geometric mean = $\sqrt[3]{5 \times 7 \times 12} = \sqrt[3]{420} = 7.49$ (2 d.p.)

You can use the key $\sqrt[\blacksquare]{\square}$ on your calculator to find the cube root.

Worked example

The geometric mean of four numbers is 6. Two of the numbers are 4.5 and 8. The third and fourth numbers are equal. Calculate the values of the third and fourth numbers. **(2)**

Geometric mean = $6 = \sqrt[4]{4.5 \times 8 \times x \times x}$

So $6^4 = 4.5 \times 8 \times x \times x$

$1296 = 36x^2$

$36 = x^2$

$\sqrt{36} = x$

The third and fourth numbers are both 6.

Use x for both the third and fourth values, because they are equal.

Raise both sides to the power 4.

Worked example

The number of cars sold by a car manufacturer increased by 4% in year 1 and decreased by 5% in year 2. Calculate the geometric mean of these two percentages. **(3)**

In year 1 the sales were multiplied by 1.04.

In year 2 the sales were multiplied by 0.95.

Geometric mean

$= \sqrt[n]{\text{value}_1 \times \text{value}_2 \times \ldots \times \text{value}_n}$

$\sqrt{1.04 \times 0.95} = 0.994$ (3 d.p.)

The sales decreased by 5%, now at 95%. This is a multiplier of $\frac{95}{100} = 0.95$

You don't need to know the actual values. Use the percentage multipliers.

Now try this

1 The geometric mean of two numbers is 2.5.
One number is increased by 12%. The other number is decreased by 15%.
Calculate the new geometric mean to 3 decimal places. **(3)**

2 The number of people aged 65 and over in the UK increased by 5.9% from 1986 to 1996, by 4.7% from 1996 to 2006 and by 22.4% from 2006 to 2016, correct to 1 decimal place.

Source: Office for National Statistics

Calculate the geometric mean of these percentage increases. **(3)**

Weighted mean

A weighted mean is one where each data value is multiplied by a number (the weight) based on importance.

The weighted mean $\bar{x}$ is given by the formula

$$\bar{x} = \frac{\sum wx}{\sum w}$$ **LEARN IT!**

where $\overline{w}$ is the weight given to each variable, x.

For example, in an interview for a job, people have to do four tasks: A, B, C and D.

The weights given to the tasks are 1, 2, 2 and 5, meaning that task D is the most important and task A the least.

Task	A	B	C	D
Weight	1	2	2	5
Jim's mark	10	8	7	4
Anne's mark	3	4	6	8

Jim's weighted mean

$$= \frac{1 \times 10 + 2 \times 8 + 2 \times 7 + 5 \times 4}{1 + 2 + 2 + 5} = 6$$

Anne's weighted mean

$$= \frac{1 \times 3 + 2 \times 4 + 2 \times 6 + 5 \times 8}{1 + 2 + 2 + 5} = 6.3$$

Worked example

tier H

In a flower show competition, displays were given a mark for shape, a mark for colour and a mark for ambience.

The table shows the weight for each quality.

Quality	Shape	Colour	Ambience
Weight	1	2	2

Mr Smith got 7 for shape, 9 for colour and 8 for ambience.
Work out his weighted mean. **(2)**

$$\text{Weighted mean} = \frac{1 \times 7 + 2 \times 9 + 2 \times 8}{1 + 2 + 2} = 8.2$$

'Ambience' and 'Colour' are each twice as important as 'Shape', as they have twice the weight.

1. Multiply each mark by its weight then add your answers.
2. Divide the total by the sum of the weights.
3. Write down all the figures on the calculator display.

Worked example

tier H

An exam consists of Paper 1 worth 80 marks and Paper 2 worth 70 marks.
The papers are equally weighted.
Liz got 52 marks for Paper 1 and 56 marks for Paper 2.
Work out the overall percentage for the exam. **(3)**

Paper 1: $\frac{52}{80} \times 100 = 65\%$

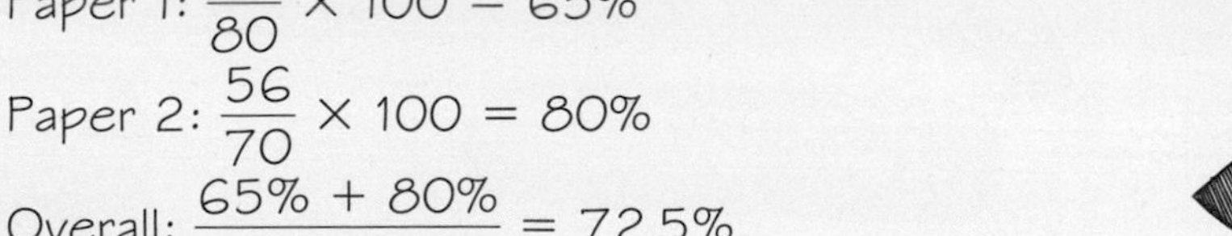

Paper 2: $\frac{56}{70} \times 100 = 80\%$

Overall: $\frac{65\% + 80\%}{2} = 72.5\%$

The first step is to work out the percentage mark on each paper. Write each mark as a fraction of the paper total and then multiply by 100.

Since the papers are equally weighted you can add the percentages together and divide by 2 as it is the same as using weights of 1.

Now try this

tier H

An exam has three papers: A, B and C.
Paper A is worth 60 marks. Paper B is worth 60 marks and Paper C is worth 80 marks. The percentage marks on the papers are equally weighted.
Ahmed got 45 on Paper A, 36 on Paper B and 60 on Paper C.
What is his mean percentage? **(3)**

You should work out the percentage mark on each paper first.

Had a look ☐ Nearly there ☐ Nailed it! ☐

Measures of dispersion for discrete data

Range and interquartile range are measures of **spread** or **dispersion**. They tell you how spread out the data is. You need to be able to calculate the **range, quartiles** and **interquartile range** of discrete data.

LEARN IT!

Quartiles divide a data set into four equal parts.

$Q_1 = \frac{n+1}{4}$th value, where n = number of data values.

Interquartile range (IQR) — Half of the values lie between the lower quartile and the upper quartile.

× × × ×× × ×××× × DATA VALUES

Smallest value | Lower quartile (Q_1) | Median (Q_2) | Upper quartile (Q_3) | Largest value

$Q_3 = \frac{3(n+1)}{4}$th value

Range = largest value − smallest value

Interquartile range = (IQR) = upper quartile (Q_3) − lower quartile (Q_1)

Worked example

tier F&H

Kim recorded the numbers of goals she scored in 12 netball games.

13 10 4 10 7 12 11 14 14 8 6 9

(a) Work out the upper and lower quartiles. **(2)**

4 6 7 8 9 10 10 11 12 13 14 14

$\frac{1}{4}(12 + 1) = 3.25$

3rd value ... 4th value

7 (7.25) 7.5 7.75 8 — Q_1 is 7.25

$\frac{3}{4}(12 + 1) = 9.75$

9th value ... 10th value

12 12.25 12.5 (12.75) 13 — Q_3 is 12.75

(b) Find the interquartile range. **(1)**

IQR = 12.75 − 7.25 = 5.5

1. Write the data in order of size.
2. Count the number of values.
3. $\frac{1}{4}(12 + 1) = 3.25$th value. Divide the interval between the 3rd and 4th values into quarters. $Q_1 = 7.25$
4. $\frac{3}{4}(12 + 1) = 9.75$th value. Divide the interval between the 9th and 10th values into quarters. $Q_3 = 12.75$
5. Find IQR = $Q_3 - Q_1$

Worked example

The table gives information about parked cars.

Number of cars	0	1	2	3	4	5
Frequency	5	16	12	10	7	3
Cumulative frequency	5	21	33	43	50	53

Find the interquartile range. **(3)**

Q_1 is the $\frac{1}{4}(53 + 1) = 13.5$th value, so $Q_1 = 1$

Q_3 is the $\frac{3}{4}(53 + 1) = 40.5$th value, so $Q_3 = 3$

IQR = 3 − 1 = 2

You **must** use cumulative frequency for any question in which you have to find the median or the interquartile range from a frequency table.

Now try this

A scientist counted the number of spots on 16 leaves of a rose bush.

3 8 0 7 4 0 8 3 2 4 3 1 1 0 2 5

(a) Work out the range. **(1)**

(b) Work out the interquartile range. **(2)**

(c) Give one advantage and one disadvantage in using the range as a measure of spread. **(2)**

Measures of dispersion for grouped data 1

When data is grouped you do not know the exact values, so you can only calculate **estimates** of the **range**, **quartiles** and **interquartile range**.

Worked example (tier F&H)

The frequency table shows the numbers of newspapers people read in one month.

Number of newspapers	Frequency
0–7	5
8–14	16
15–21	22
22–50	12

Calculate an estimate for the range. **(1)**

Range = largest value − smallest value

Smallest possible value is 0, largest possible value is 50.

Range = 50 − 0 = 50

Worked example (tier F&H)

The table shows the weights, w, to the nearest kg, of students in a drama group.

Weight, w (kg)	Frequency
$45 < w \leqslant 50$	10
$50 < w \leqslant 55$	12
$55 < w \leqslant 60$	15
$60 < w \leqslant 65$	8

Estimate the range of weights. **(1)**

Range = 65.5 − 45.5 = 20 kg

When data is rounded think carefully about the minimum and maximum possible values.
$w > 45$ so minimum possible weight is 45.5 kg.
$w \leqslant 65$ so maximum possible weight is 65.5 kg.

Worked example (tier F&H)

Distance, x (km)	Frequency	Cumulative frequency
$0 < x \leqslant 5$	34	34
$5 < x \leqslant 10$	38	72
$10 < x \leqslant 15$	22	94
$15 < x \leqslant 20$	16	110
$20 < x \leqslant 25$	10	120

The grouped frequency table gives data about the distances 120 people travelled to work. Estimate the three quartiles. **(4)**

$Q_1 = \frac{120}{4}$
= 30th value
≈ 4.5 km

$Q_2 = \frac{120}{2}$
= 60th value
≈ 9.5 km

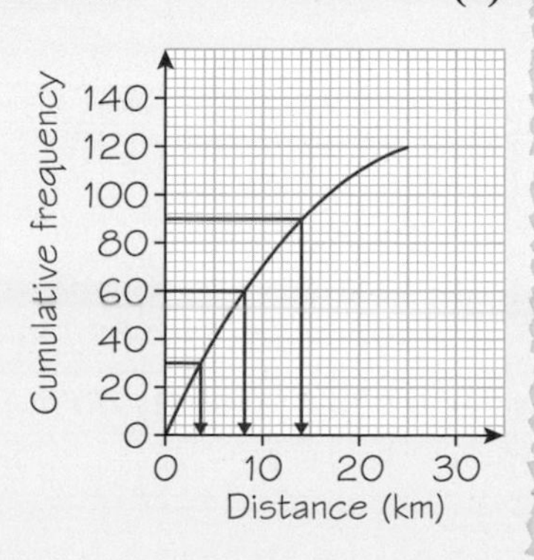

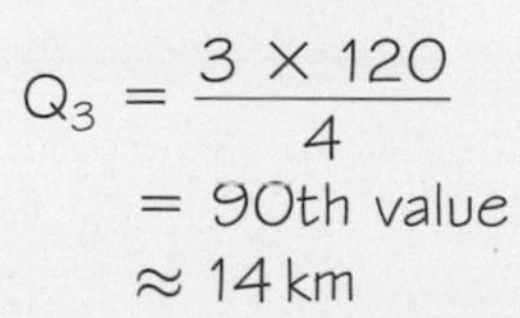

$Q_3 = \frac{3 \times 120}{4}$
= 90th value
≈ 14 km

You can also use the table and linear interpolation.

$Q_1 = 0 + \frac{30}{34} \times 5 = 4.41$

Now try this (tier F&H)

The table gives information about the times, in seconds, people could read without blinking.

Time, t (seconds)	Frequency
$0 < t \leqslant 10$	4
$10 < t \leqslant 20$	8
$20 < t \leqslant 30$	11
$30 < t \leqslant 40$	15
$40 < t \leqslant 50$	14
$50 < t \leqslant 60$	4

Find an estimate for the interquartile range:

(a) using a cumulative frequency graph **(4)**

(b) using linear interpolation. **(4)**

Had a look ☐ Nearly there ☐ Nailed it! ☐

Measures of dispersion for grouped data 2

Deciles divide data into 10 equal parts. **Percentiles** divide data into 100 equal parts. Deciles and percentiles of continuous data can be found using cumulative frequency diagrams or linear interpolation.

Finding deciles

To find the dth decile read across from $\frac{d}{10} \times n$ on the vertical axis of a cumulative frequency diagram.

You may be asked to find an **interdecile range**. For example, the 4th–6th interdecile range is the 6th decile minus the 4th decile. It uses the middle 20% of the data.

LEARN IT!

Finding percentiles

To find the pth percentile read across from $\frac{p}{100} \times n$ on the vertical axis of a cumulative frequency diagram.

You may be asked to find an **interpercentile range**. For example, the 10th–90th interpercentile range is the 90th percentile minus the 10th percentile. It uses the middle 80% of the data.

This cumulative frequency diagram shows how late 50 people were for work one Monday.

To find the 76th percentile read across from $\frac{76}{100} \times 50 = 38$. The 76th percentile is 19 minutes.

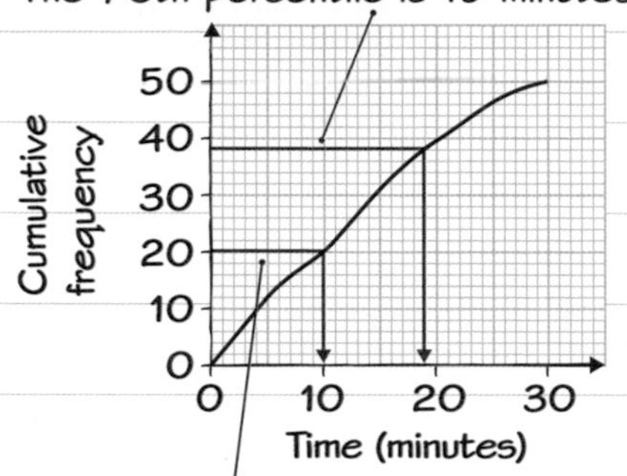

To find the 4th decile read across from $\frac{4}{10} \times 50 = 20$. The 4th decile is 10 minutes.

Worked example

tier H

The cumulative frequency diagram gives data about the distances 150 people travelled to work.

(a) Find an estimate for the 6th decile. **(1)**

6th decile: $\frac{6}{10} \times 150 = 90\text{th value} \approx 11\text{ km}$

(b) Find an estimate for the 36th percentile. **(2)**

36th percentile: $\frac{36}{100} \times 150 = 54\text{th value} \approx 7\text{ km}$

(c) Work out the 20th to 80th interpercentile range. **(2)**

$18 - 5 = 13\text{ km}$

Work out the position of the 6th decile up the vertical axis: $\frac{6}{10} \times 150$

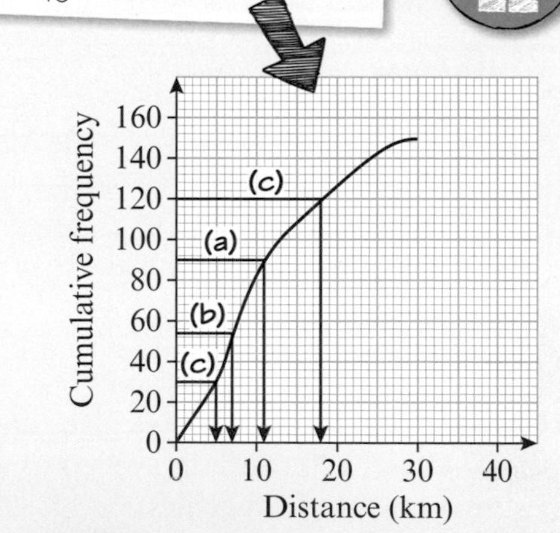

Find the 80th percentile and the 20th percentile. Work out 80th percentile – 20th percentile.

Now try this

tier H

The frequency table gives information about the times, in seconds, people can balance on one leg.

Time, t (seconds)	$0 < t \leqslant 10$	$10 < t \leqslant 20$	$20 < t \leqslant 30$	$30 < t \leqslant 40$	$40 < t \leqslant 50$	$50 < t \leqslant 60$
Frequency	5	9	10	14	20	2

Use the table to find estimates for the value of:

(a) the 20th percentile **(1)**

(b) the 6th to 9th interdecile range. **(2)**

Standard deviation 1

The **standard deviation (SD)** of a distribution is a measure of how much all the values deviate from the mean value, or how spread out they are. The deviation or dispersion of an observation, x, from the mean $\bar{x}$ is given by $x - \bar{x}$.

There are two methods to work out the standard deviation of a set of data values using these formulae.

1 Standard deviation $= \sqrt{\frac{1}{n}\sum(x - \bar{x})^2}$

or

2 Standard deviation $= \sqrt{\frac{\sum x^2}{n} - \left(\frac{\sum x}{n}\right)^2}$

The closer the xs are to $\bar{x}$, the smaller $\sum(x - \bar{x})^2$ is, and the smaller the standard deviation.

Worked example

tier H

Find the standard deviation for these values.

70 100 130 80 90 40 60 70 **(2)**

Method 1

Standard deviation $= \sqrt{\frac{1}{n}\sum(x - \bar{x})^2}$

Calculate the mean, $\bar{x}$.

$$\bar{x} = \frac{70 + 100 + 130 + 80 + 90 + 40 + 60 + 70}{8}$$

$$= 80$$

$$\sum(x - \bar{x})^2 = 100 + 400 + 2500 + 0 + 100 + 1600 + 400 + 100$$
$$= 5200$$

Calculate the deviation of each data value from the mean $(x - \bar{x})$ and square it $(x - \bar{x})^2$. For the first value, $70 - 80 = -10$, $(-10)^2 = 100$, and so on. Add them all together.

$$\frac{1}{n}\sum(x - \bar{x})^2 = \frac{5200}{8}$$

$$\sqrt{\frac{1}{n}\sum(x - \bar{x})^2} = \sqrt{\frac{5200}{8}} = 25.49... = 25.5 \text{ (3 s.f.)}$$

Method 2

Standard deviation $= \sqrt{\frac{\sum x^2}{n} - \left(\frac{\sum x}{n}\right)^2}$

You can describe $\sqrt{\frac{\sum x^2}{n} - \left(\frac{\sum x}{n}\right)^2}$ as 'the square root of (the mean of the squares minus the square of the mean)'.

$$\text{Mean} = \frac{\sum x}{n} = \frac{640}{8} = 80$$

Calculate the mean.

$$\sum x^2 = 70^2 + 100^2 + 130^2 + 80^2 + 90^2 + 40^2 + 60^2 + 70^2$$
$$= 56400$$

$$\text{Standard deviation} = \sqrt{\frac{56400}{8} - 80^2}$$
$$= \sqrt{650}$$
$$= 25.5 \text{ (3 s.f.)}$$

Square every value and find the total.

Now try this

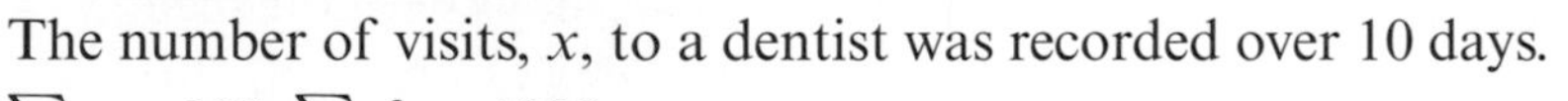

The number of visits, x, to a dentist was recorded over 10 days.

$\sum x = 200$, $\sum x^2 = 4800$

Work out the mean and the standard deviation. **(3)**

Had a look ☐ Nearly there ☐ Nailed it! ☐

Standard deviation 2

Calculating standard deviation for a frequency table or grouped data

Two formulae can be used to calculate the standard deviation for a **frequency table** or **grouped data**.

$$\text{Standard deviation} = \sqrt{\frac{\sum f(x-\bar{x})^2}{\sum f}} = \sqrt{\frac{\sum fx^2}{\sum f} - \left(\frac{\sum fx}{\sum f}\right)^2}$$

Remember that in a frequency table, the total frequency $\sum f = n$ and the mean is $\frac{\sum fx}{\sum f}$.

See pages 42–44 for more on averages from grouped data.

These formulae are **not** on the formula sheet in the exam paper.

Both versions of the formula can be used to work out the standard deviation for a frequency table of **discrete** data.

Worked example

tier H

The table shows the numbers of goals scored by a football team in 25 games.
Calculate the standard deviation. **(4)**

Number of goals	0	1	2	3	4
Frequency	2	8	9	5	1
fx	0	8	18	15	4
$x - \bar{x}$	−1.8	−0.8	0.2	1.2	2.2
$(x - \bar{x})^2$	3.24	0.64	0.04	1.44	4.84
$f(x - \bar{x})^2$	6.48	5.12	0.36	7.2	4.84

Calculate the mean, $\bar{x}$.

$$\bar{x} = \frac{0 \times 2 + 1 \times 8 + 2 \times 9 + 3 \times 5 + 4 \times 1}{25} = 1.8$$

Calculate the value of $\sum f(x - \bar{x})^2$.

$$\sum f(x - \bar{x})^2 = 24$$

Substitute the values for $\sum f(x - \bar{x})^2$ and $\sum f$.

$$\sqrt{\frac{\sum f(x-\bar{x})^2}{\sum f}} = \sqrt{\frac{24}{25}} = 0.980 \text{ (3 s.f.)}$$

Worked example

In one week 115 students were late for school. The school recorded the number of times, x, each student was late to school in this week.
Given that $\sum fx = 275$ and $\sum fx^2 = 821$

(a) work out the mean number of times a student was late for school **(1)**

$$\text{Mean} = \frac{\sum fx}{\sum f} = \frac{275}{115} = 2.4 \text{ times}$$

(b) find the standard deviation to one decimal place. **(3)**

$$\text{SD} = \sqrt{\frac{\sum fx^2}{\sum f} - \left(\frac{\sum fx}{\sum f}\right)^2} = \sqrt{\frac{821}{115} - (2.4)^2}$$
$$= 1.17$$

The number of children, x, seen at a surgery on 20 days has the following data:

$\bar{x} = 10, \sum fx^2 = 2320$

(a) Work out the standard deviation. **(2)**

On each of days 21 and 22, 10 children were seen at the surgery.

(b) Compare the mean and the standard deviation for all 22 days with the mean and the standard deviation for the first 20 days. **(2)**

Standard deviation 3

For **grouped continuous data** the easiest formula to use for the standard deviation is:

$$\text{Standard deviation} = \sqrt{\frac{\sum fx^2}{\sum f} - \left(\frac{\sum fx}{\sum f}\right)^2}$$

If the data is continuous then the formula gives an estimate for the standard deviation – just as the formula on page 43 for the mean gave an estimate for the mean of continuous data.

The table gives information about the times (in seconds) it took some students to solve a puzzle.

The three columns in red are working columns you should add to the table to find the values you will need to calculate the standard deviation:

Time, T (s)	Frequency, f	Mid T	fT	fT^2
$0 < T \leqslant 20$	6	10	60	600
$20 < T \leqslant 40$	11	30	330	9900
$40 < T \leqslant 60$	13	50	650	32 500
$60 < T \leqslant 80$	6	70	420	29 400
$80 < T \leqslant 100$	4	90	360	32 400
	40		1820	104 800

In most questions you will not need to calculate every value in a table.

$\sum f = 40$ $\sum fT = 1820$ $\sum fT^2 = 104\,800$

Mean $= 1820 \div 40 = 45.5$

Standard deviation $= \sqrt{104\,800 \div 40 - 45.5^2} = 23.4$ (3 s.f.)

Worked example

tier H

The partially completed table gives data about the distances that some birds flew from their nests and back.

Distance, D (m)	Frequency, f	Mid D	fD	fD^2
$0 < D \leqslant 20$	19	10	190	1900
$20 < D \leqslant 40$	14	30	420	12 600
$40 < D \leqslant 60$	10	50	500	25 000
$60 < D \leqslant 80$	10	70	700	49 000
$80 < D \leqslant 100$	7	90	630	56 700
	60		2440	145 200

fD^2 means $f \times D^2$ so $14 \times 30^2 = 14 \times 900$

Put the sums of the f, the fD and the fD^2 columns at the bottom of the table.

(a) Complete the table. **(1)**

(b) Estimate the standard deviation. **(2)**

$\sum f = 60$ $\sum fD = 2440$ $\sum fD^2 = 145\,200$

Mean $= 2440 \div 60 = 40.66667$ m

Standard deviation $= \sqrt{145\,200 \div 60 - 40.66667^2}$

$= 27.7$ (3 s.f.)

Work out the mean. Then use the formula for standard deviation.

Now try this

tier H

1 The times, t seconds, that a group of 30 people took to solve a puzzle are summarised by the following data: $\sum ft = 1200$, $\sum ft^2 = 51\,000$. Find estimates for the mean and the standard deviation. **(3)**

2 The table shows the heights, in centimetres, of 100 students.

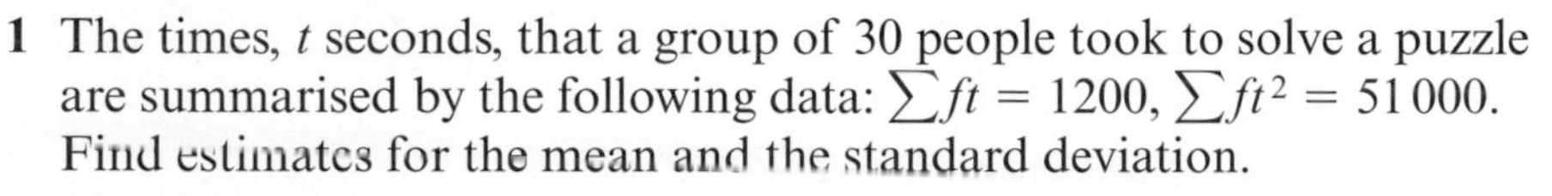

Height, h (cm)	$120 \leqslant h < 130$	$130 \leqslant h < 140$	$140 \leqslant h < 150$	$150 \leqslant h < 160$	$160 \leqslant h < 170$
Frequency	9	17	23	32	19

(a) Work out an estimate of the mean height. **(3)**

(b) Work out an estimate of the standard deviation. **(2)**

(c) Explain why your values for the mean and standard deviation are estimates. **(1)**

Box plots

Box plots are used to display the **minimum value**, the **maximum value**, the **lower quartile**, the **upper quartile** and the **median** from a distribution.

If you are asked to draw a box plot these are the five pieces of information you must put on the grid.

The lowest 25% of the values are less than or equal to the lower quartile, Q_1.

The highest 25% of the values are greater than or equal to the upper quartile, Q_3.

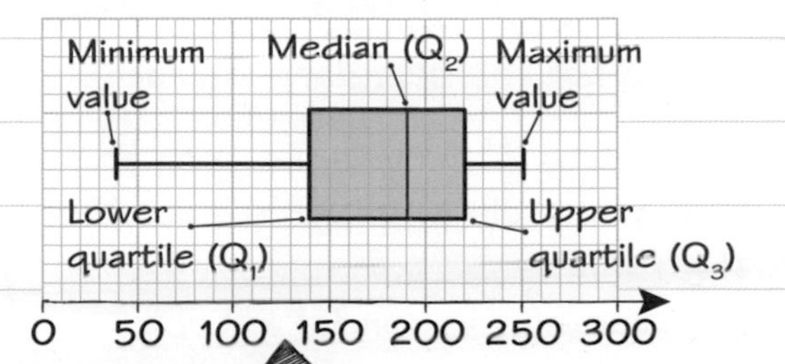

A box plot is always drawn on graph paper and always includes a scale.

Interpreting box plots

The box plot shows information about the total numbers of passengers on different bus routes one morning.

The lowest number of passengers was 40.

The median was 160.

Q_1 = 140 so 25% of the bus routes had 140 passengers or fewer.

Q_3 = 220 so 25% of the bus routes had 220 passengers or more.

The interquartile range was $Q_3 - Q_1 = 220 - 140 = 80$

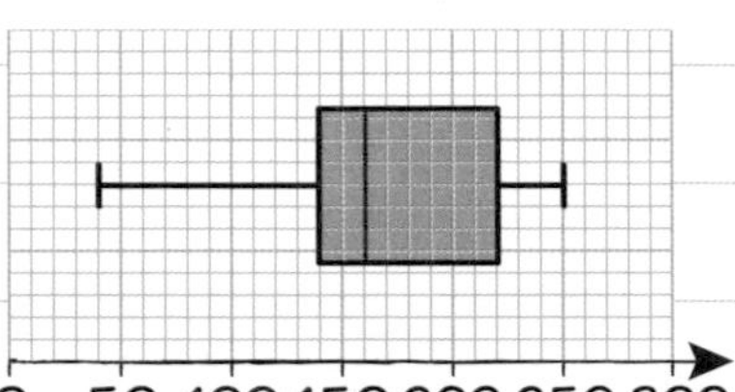

Worked example

tier F&H

Here are the ages of some people in a club.

10 11 11 12 15 18 18 19 23 29
35 36 41 41 48

Draw a box plot to show this information. **(3)**

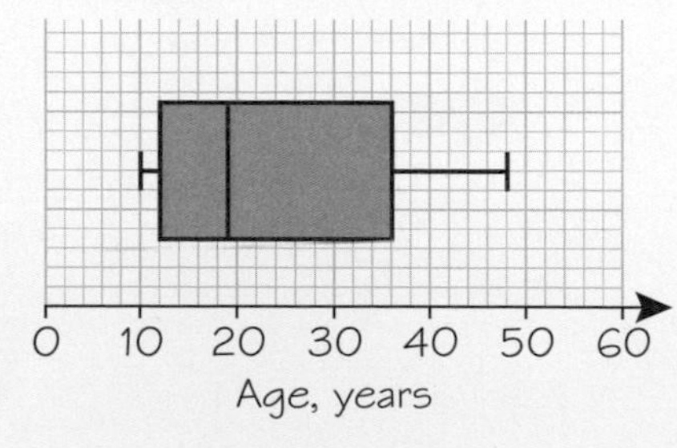

The ages are already in order. If they were not, the first step would be to put them in order.

1. The lowest value is 10 and the highest value is 48.
2. Draw the scale from 0 to 60 and mark the lowest and highest values.
3. Count the number of values, n.
4. $\frac{n+1}{2} = 8$ so the median is the 8th value (19).
5. $\frac{n+1}{4} = 4$ so Q_1 is the 4th value (12).
6. $\frac{3(n+1)}{4} = 12$ so Q_3 is the 12th value (36).
7. Complete the diagram by marking Q_1, the median and Q_3.

Now try this

tier F&H

The list shows some information about the numbers of people attending a community centre over 15 nights.

14 17 43 25 12 25 34 43 28 19 15 23 24 34 30

Draw a box plot to show this information. **(3)**

Start by putting the data values in order, smallest first.

Outliers

An extreme value in a distribution is called an **outlier**.

Outliers on box plots

LEARN IT!

An outlier is any data value that is

- less than $Q_1 - 1.5 \times \text{IQR}$
- greater than $Q_3 + 1.5 \times \text{IQR}$

You can represent an outlier with a cross (×) on a box plot. Your 'whiskers' should only extend as far as the lowest and highest data values that are **not** outliers.

This data shows the numbers of times 11 friends checked social media in one day. The upper and lower quartiles are circled.

3 10 (14) 15 16 16 18 18 (19) 20 21

$\text{IQR} = Q_3 - Q_1 = 19 - 14 = 5$

The smallest data value that is **not** an outlier is 10, so draw the end of this whisker at 10.

The highest data value is not an outlier, so draw this whisker at 21.

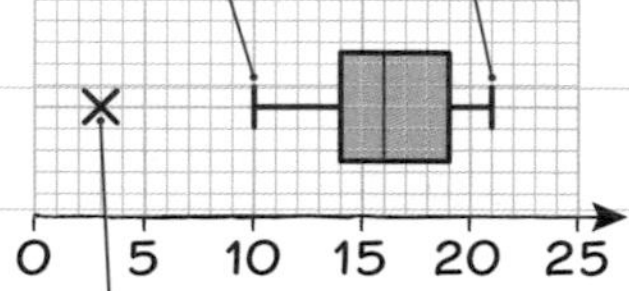

Any outliers are shown as crosses.

$Q_1 - 1.5 \times \text{IQR} = 14 - 1.5 \times 5 = 6.5$ so any data values less than 6.5 are outliers. The only outlier is 3.

Outliers in standard deviation calculations

LEARN IT!

An outlier is defined as a value more than 3 standard deviations from the mean. An outlier is outside $\bar{x} \pm 3\sigma$ where $\bar{x}$ = mean and σ = standard deviation.

In a larger survey of social media use, the mean number was 18 with standard deviation 3.4.

The outliers are:

$18 - 3 \times 3.4 = 7.8$ so fewer than 8 times.

$18 + 3 \times 3.4 = 28.2$ so more than 28 times.

Worked example

tier H

The list shows the numbers of times people accessed their email account, in one day.

0 0 1 3 3 4 5 6 7 7 9 9 10 15 19

Identify any values which are outliers. **(3)**

Q_1 is the 4th value, $Q_1 = 3$

Q_3 is the 12th value, $Q_3 = 9$

$\text{IQR} = 9 - 3 = 6$ $\quad 1.5 \times 6 = 9$

$Q_3 + 9 = 18$, so 19 is an outlier.

$Q_1 - 9 = -6$ so there are no low outliers.

Worked example

In a maths test the mean mark was 46% and the standard deviation was 10.8.

Amy scored 95% on the test.

Show that her result is an outlier. **(2)**

$46 + 3 \times 10.8 = 78.4\%$

95% > 78.4% so Amy's result is an outlier.

Now try this

1 The list shows information about the numbers of people attending a keep-fit class over 11 days.

4 5 8 9 10 13 16 17 17 20 24

(a) Work out the values of the quartiles. **(1)**

(b) Identify any values which are outliers. **(2)**

2 In an experiment comparing the weights of tomatoes produced by different types of plants in one season, the mean weight of tomatoes produced was 9.05 pounds with standard deviation 1.4. One variety of plant produced 4.6 pounds. Show that this weight is an outlier. **(2)**

Had a look ☐ Nearly there ☐ Nailed it! ☐

Skewness

Skewness describes the distribution of data compared with the averages of the data. You need to be able to determine skewness from data by inspection and by calculation.

Distribution of data can be **symmetrical**, or have **positive skew** or **negative skew**. You can also identify skewness from box plots by inspection.

Positive skew

The median lies towards the left-hand end of the box plot.

(median − LQ) < (UQ − median)

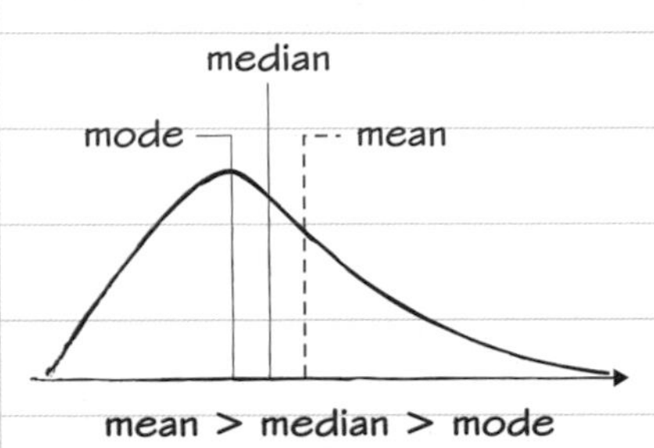

Symmetrical distribution

The median lies in the middle of the box plot.

(median − LQ) = (UQ − median)

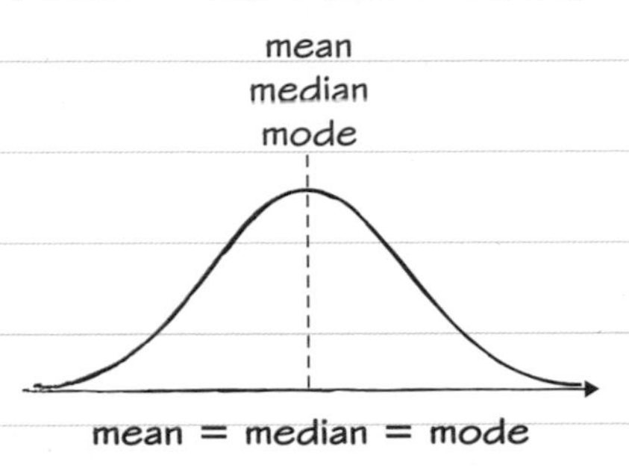

Negative skew

The median lies towards the right-hand end of the box plot.

(UQ − median) < (median − LQ)

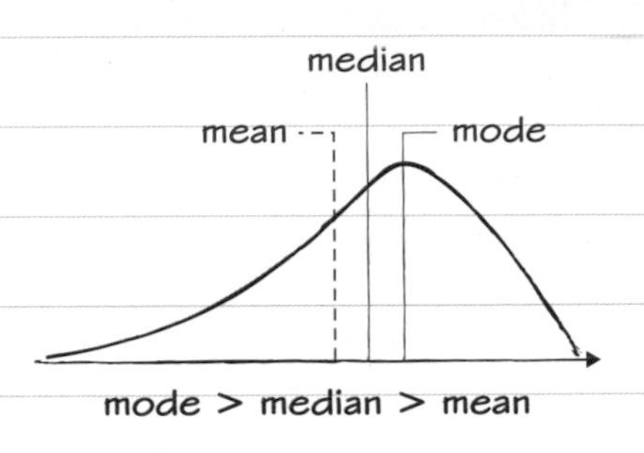

Worked example (tier F&H)

The list shows the numbers of times some people used their mobile phones in one day.

2 4 7 10 0 5 4 1 12 14 12 4 18 15 6

(a) Draw a box plot to show this information. **(3)**

0 1 2 (4) 4 4 5 (6) 7 10 12 (12) 14 15 18

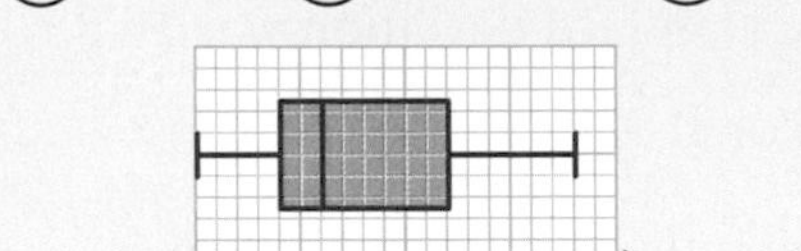

Number of times

(b) Describe the skew of the data. **(2)**

The median is closer to the LQ.

(6 − 4) < (12 − 6) i.e. 2 < 6 so positive skew

(c) Calculate the mean of the data. **(2)**

$$\text{Mean} = \frac{114}{15} = 7.6 \text{ times}$$

(d) The standard deviation for this data is 5.54. Work out the value of the skew to 2 d.p. and interpret your result. **(2)**

$$\text{Skew} = \frac{3(7.6 - 6)}{5.54} = 0.87$$

This implies a strong positive skew.

Calculating a value for skew

$$\text{Skew} = \frac{3(\text{mean} - \text{median})}{\text{standard deviation}}$$

The skew value can be:

- positive: indicates positive skew, the higher the value the stronger the skew
- negative: indicates negative skew, the lower the value the stronger the skew
- zero: indicates the distribution is symmetrical.

Now try this (tier F&H)

The list shows the numbers of people attending an art club for 11 days.

4 6 8 9 11 13 15 17 17 21 24

(a) Draw a box plot. **(2)**

(b) Describe and interpret the skew of the data. **(2)**

(c) Calculate the mean of the data. **(2)**

(d) Compare the mean, median and mode. **(1)**

(e) Given the standard deviation for this data is 6.49, calculate the skew of this distribution and interpret your result. **(2)**

Deciding which average to use

Average	Advantages	Disadvantages
Mode	• Easy to find and always a data value • Can be used with any type of data • Unaffected by open-ended or extreme values	• May not exist • Cannot be used to calculate a measure of spread
Median	• Unaffected by extreme values • Can be used to help calculate quartiles, interquartile range and skew	• May not be a data value
Mean	• Uses all the data • Can be used to calculate standard deviation and skew	• Always affected by open-ended or extreme values • Rarely a data value

Worked example

The stem and leaf diagram shows information about the number of patients in A&E on each of 15 nights.

```
1 | 2  2  4
2 | 0  3  6  7
3 | 1  2  3  6  7
4 | 0  2  3
```

Key 4|3 represents 43 patients

(a) Find the median. **(1)**

$\frac{n+1}{2} = \frac{16}{2} = 8$ The 8th value is 31.

On the next two nights the numbers of patients were 38 and 40.

(b) What effect does this have on the median? **(1)**

$\frac{n+1}{2} = \frac{18}{2} = 9$ The median increases to 32.

Worked example

The mean of 10 tests for Jack is 17.5.
Each test is marked out of 20.
Can Jack get enough marks in his next test to bring his mean mark up to 18? **(2)**

Total so far = 10 × 17.5 = 175

Desired total = 11 × 18 = 198

198 − 175 = 23 so it's not possible.

You can multiply the mean by the number of data values to work out the total of all the data values.

Now try this

1 The table shows the numbers of computers in a sample of people's houses.

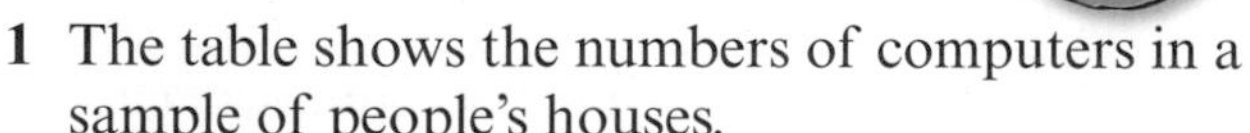

Number	0	1	2	3	4	5
Frequency	4	7	9	6	4	1

(a) Work out the median. **(2)**

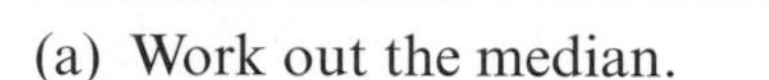

(b) Work out the mean. **(2)**

An additional house has 15 computers.

(c) Which of the two averages will be most affected if this house is included in the sample? **(2)**

tier H

2 These are the results of a survey on the numbers of text messages students sent in 1 day.

Number of texts	0–2	3–19	20–50	51–70
Percentage	3	38	53	6

Explain which type of average should be used to represent this data. **(2)**

Had a look ☐ Nearly there ☐ Nailed it! ☐

Comparing data sets

Comparing lists

Here is a list of marks for the boys and girls in Mrs Watson's class.

Boys	6	7	12	13	16	16	18	20
Girls	8	8	9	13	14	15	20	

To compare the distributions, first of all use the median or the mean.

Then work out the range for each list.

Boys' median $= \frac{13 + 16}{2} = 14.5$

Girls' median $= 13$

In general the boys had higher marks than the girls because they had a higher median.

Boys' range $= 20 - 6 = 14$

Girls' range $= 20 - 8 = 12$

The spread of the boys' marks was greater than the girls' because the boys had a greater range.

Golden rules

When comparing data:

- always work out an average and make a comment
- always work out a measure of spread and make a comment.

If you work out the

- mode, then compare the range
- median, then compare the range or interquartile range
- mean, then compare the range or standard deviation.

You can also compare the skew of the distributions.

Worked example

Here is a list of the numbers of cars arriving at a crossroads each minute for 15 minutes.

2 2 3 5 5 6 6 6 7 7 7 8 9 9 11

The box plot shows information about the numbers of vans arriving at the crossroads each minute.

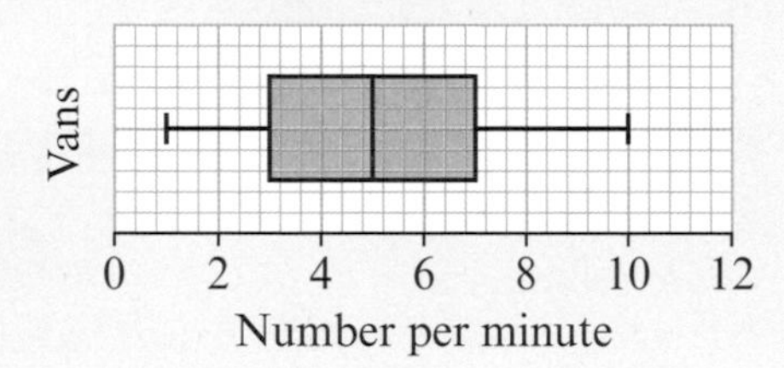

(a) Work out the median for the cars. **(1)**

Median $= \frac{15 + 1}{2}$ th value $= 6$

(b) Work out the lower quartile for the cars. **(1)**

Lower quartile $= \frac{15 + 1}{4}$th value $=$ 4th number in the list $= 5$

(c) Compare the distribution of the cars with the distribution of the vans. **(2)**

The median for the cars is 6 and for the vans is 5.

On average more cars arrived each minute. The interquartile range for the cars is $8 - 5 = 3$ and for the vans is $7 - 3 = 4$ showing that the distribution of the cars is more concentrated about the median.

Compare medians from your calculation and from the box plot.

Calculate the IQR from the list and compare this with the value found from the box plot. You could also compare the ranges or the skews of the distributions.

Now try this

1 Here is a list of the numbers of shops the people in group A visited in one week.

5 8 6 7 9 1 4 6 2 12 10 3 0 8 13

For the people in group B, the median was 7.5 and the interquartile range was 5.

Compare the two groups, A and B. **(3)**

2 The table shows summary statistics for the lengths of a sample of American cars and European cars.

tier H

	Mean length (cm)	Standard deviation
American cars	498	50.8
European cars	425	34.3

Compare the distribution of the lengths of the American cars and the European cars. **(2)**

Making estimates

You can use the mean, median, range and interquartile range of a **representative sample** to **estimate** the same statistics for the **whole population**.

In a distribution:

- 50% of the data in a distribution is less than the median, and 50% is greater than the median
- 25% of the data is less than the lower quartile
- 25% of the data is greater than the upper quartile
- 50% of the data is between the lower and upper quartiles.

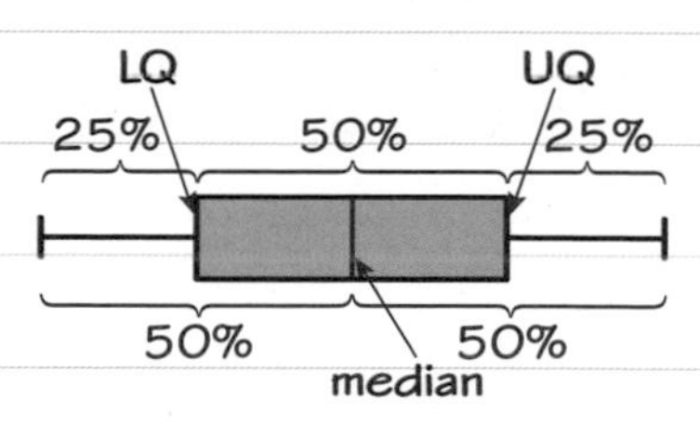

Worked example

tier F&H

The systolic blood pressures of a representative sample of patients admitted to a hospital in one day were recorded.
The table shows the results.

Minimum	Lower quartile	Median	Upper quartile	Maximum
107.5	112.4	116.7	124.9	160.1

(a) What proportion of the sample had a systolic blood pressure less than 112.4? **(1)**

25% of the sample had a systolic blood pressure less than 112.4

The total number of patients admitted to hospital in the UK in one day was 41 500.

(b) Estimate the number of patients admitted in the UK who

(i) had a systolic blood pressure less than 112.4 **(2)**

25% of 41500 = $0.25 \times 41500 = 10375$

(ii) had a systolic blood pressure greater than 116.7. **(2)**

50% of 41500 = $0.5 \times 41500 = 20750$

112.4 is the lower quartile.

116.7 is the median. 50% have a systolic blood pressure greater than the median.

Now try this

tier F&H

1 The table shows the results of a survey on the weights, in kg, of a random sample of 18-month-old boys.

Percentile	5	10	15	25	50	75	85	90	95
Weight (kg)	8.9	9.2	9.8	10.5	11.5	12.6	13.3	13.8	14.4

The number of 18-month-old boys in Manchester one year was 4115.

How many of these boys would you expect to weigh

(a) 9.8 kg or less **(2)**

(b) between 10.5 kg and 13.3 kg? **(3)**

2 The table shows information about the total income before tax (in £ thousands) of the 30.7 million tax payers in the UK in the tax year 2014–2015. *Source: Office for National Statistics*

Percentile	20	25	40	50	60	75	90
Income (£000s)	14.4	15.5	19.3	22.4	26.2	34.5	51.4

(a) What percentage of the tax payers had an income before tax of between £15 500 and £26 200? **(2)**

(b) How many of the tax payers would you expect to have an income before tax of £34 500 or more? **(3)**

 Had a look ☐ Nearly there ☐ Nailed it! ☐

Scatter diagrams

Scatter diagrams are used to show whether two sets of data are **associated**. This means there is a **relationship** between them.

Types of variables

Scatter diagrams are useful to represent **bivariate data**. They are used to investigate how changing one variable affects another variable. The **explanatory (independent) variable** (the one you change) is plotted on the horizontal (x) axis. The **response (dependent) variable** (the one that responds to, or depends on, the explanatory variable) is plotted on the vertical (y) axis.

The height of a plant is the response or dependent variable.

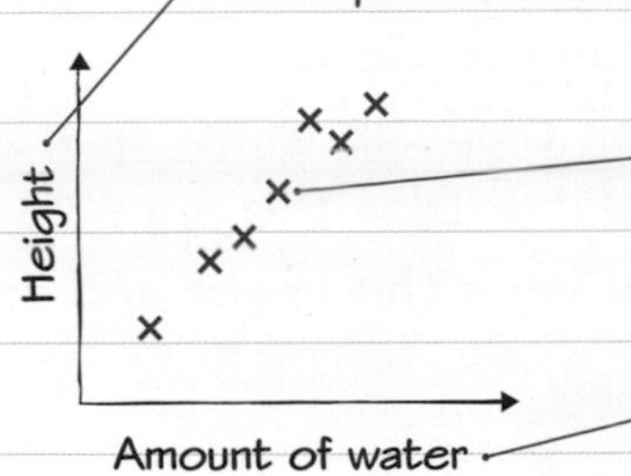

Each point corresponds to the amount of water and the height for one plant.

The amount of water is the explanatoryor independent variable.

This scatter diagram shows the heights of some plants when given different amounts of water. It shows an association between the variables. As the amount of water increases, the height increases.

Worked example

tier F

Katie records the numbers of hours that 10 students spend revising for an exam and the marks they achieve in the exam.

Student	A	B	C	D	E	F	G	H	I	J
Time	5	7	9	4	6	2	7	5	1	8
Mark (%)	80	75	92	60	70	40	84	68	35	95

Mark is the response variable.
Hours is the explanatory variable.

(a) Which variable should she plot on the horizontal axis? Give a reason for your answer. **(2)**

She should plot hours on the horizontal axis because it is the explanatory variable.

(b) Plot this data on a scatter diagram. **(3)**

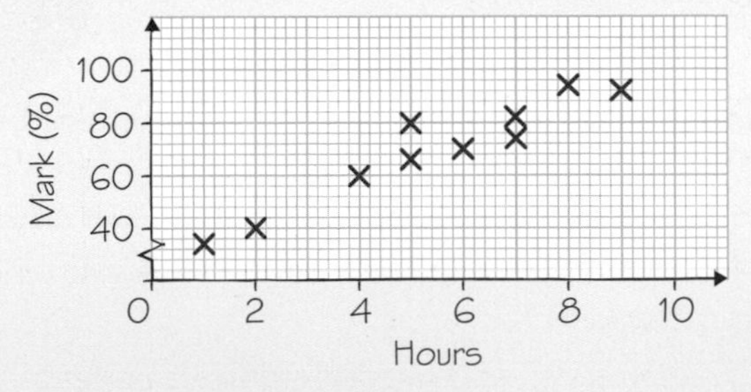

(c) Is there an association between the time students spent revising and their mark in the exam? **(1)**

Generally students who spent more hours revising achieved higher marks so there is an association between the two sets of data.

Does an increase in one set of data cause an increase or decrease in the other? The plotted points lie close to a straight line so the association is strong.

Now try this

Abi is investigating whether there is a relationship between temperature and rainfall. She collected data and plotted it on a scatter diagram.

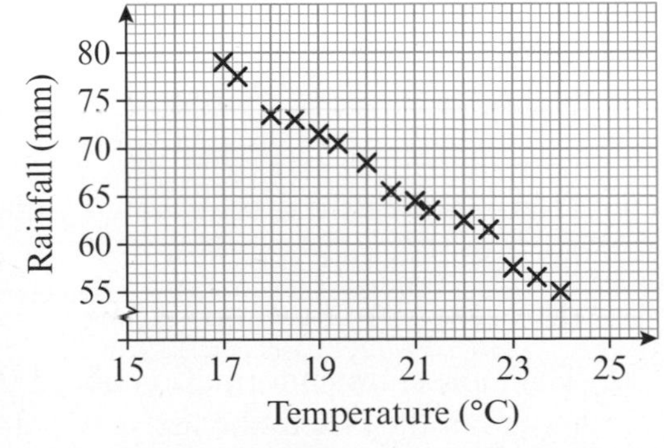

(a) Discuss whether or not this is a good diagram to use. **(2)**

(b) What does the scatter diagram tell you about the temperature in °C and the rainfall in mm? **(1)**

Correlation

A scatter diagram can be used to identify a **correlation** between two variables. Correlation describes the association, which can show an increasing or decreasing **trend**. Correlation can be strong or weak.

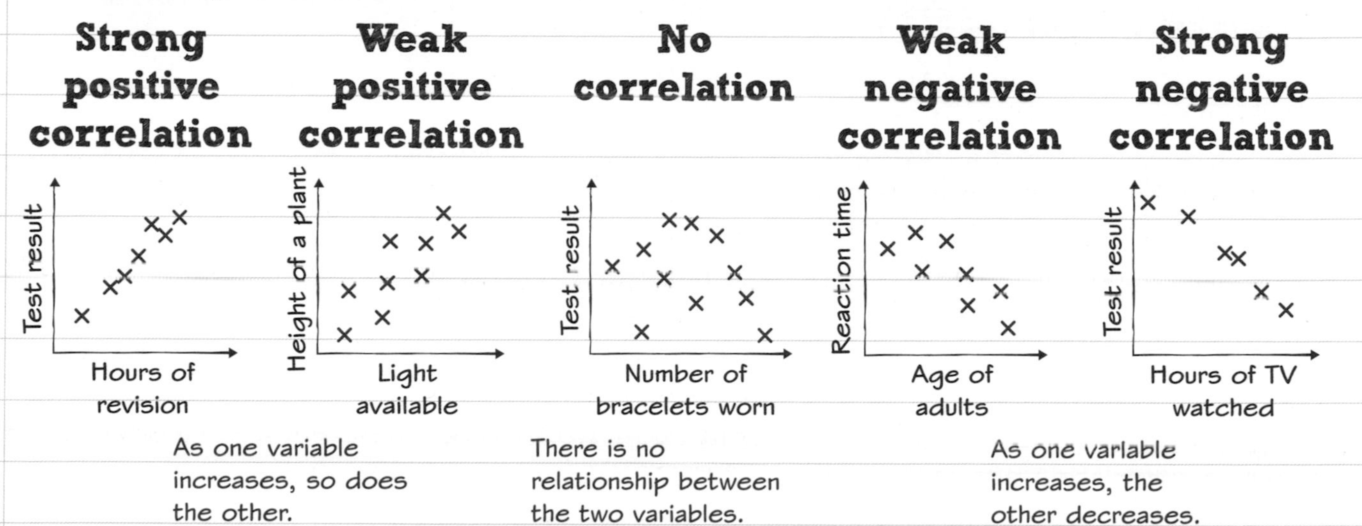

As one variable increases, so does the other.

There is no relationship between the two variables.

As one variable increases, the other decreases.

This scatter diagram is an example of positive **non-linear correlation**. It shows the relationship between the radius of a circle and its area. There is an increasing trend and the pattern formed by the points is a curved line.

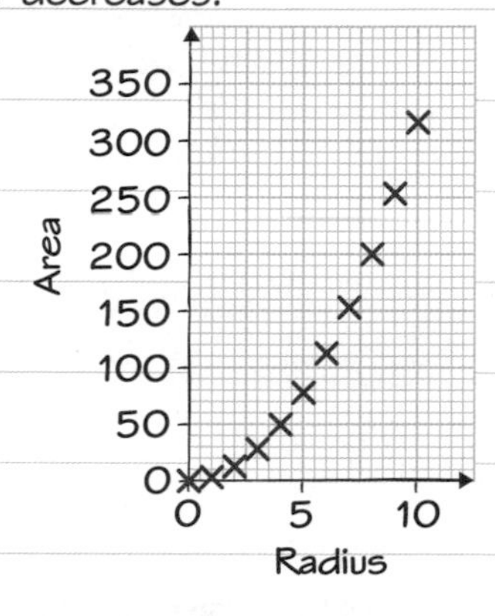

Worked example

tier F

The table gives information about the average monthly temperature, x (°C), and the money made from coat sales, y (£000s), in a shop.

Temperature, x (°C)	4	3	8	12	18	25	22
Money made, y (£1000s)	46	50	49	23	14	4	5

(a) Plot a scatter diagram to show this information. **(2)**

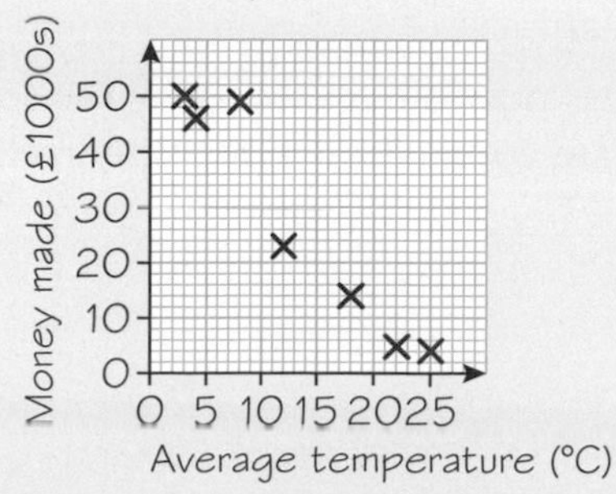

(b) Describe and interpret the type of correlation shown by the scatter diagram. **(3)**

Strong negative correlation. As the temperature increases, the money made from coat sales decreases.

Write down the type and strength of correlation, and answer the question in context. Make sure you say how temperature affects the amount of money made.

Now try this

tier F

For each of the following, identify the explanatory variable and the response variable, and state what type of correlation you expect.

(a) The weight of a child and his or her age. **(2)**

(b) The length of a train journey and the time it takes. **(2)**

(c) The average daily temperature and the number of cups of hot tea drunk. **(2)**

(d) The number of bushes in a garden and the number of people who live in the house. **(2)**

Had a look ☐ Nearly there ☐ Nailed it! ☐

Causal relationships

A **causal relationship** between two variables means that a change in one of the variables directly causes a change in the other variable, such as the number of hours of sunshine and the temperature, or the age of a car and its value.

Correlation and causation

Correlation between two variables shows that there may be a relationship between them, but it does **not** imply that a change in one variable caused the change in the other. You need to know that correlation **does not** necessarily imply **causation**.

These scatter diagrams show **causation** between temperature and the number of sales of sunglasses and between temperature and ice-cream sales.

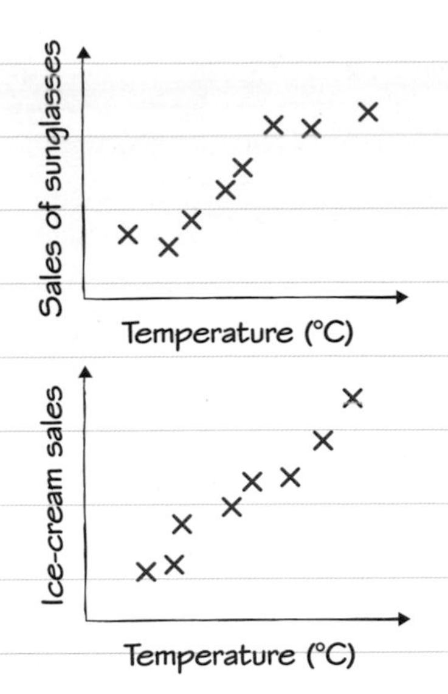

As ice-cream sales and the number of sales of sunglasses both increase as temperature increases, a scatter diagram of ice-cream sales against number of sales of sunglasses would show a positive correlation between them.

However, there is no causal relationship: an increase in ice-cream sales **does not** lead to an increase in the number of sales of sunglasses. They are both causally related to temperature, but not to each other.

Worked example

tier F&H

The scatter diagram shows the numbers of umbrellas sold and the numbers of road accidents in a town over a 12-month period. There is a positive linear correlation between the number of umbrellas sold and the number of road accidents.

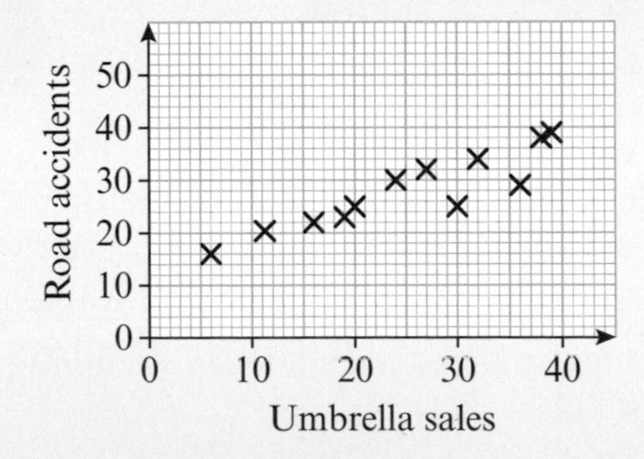

David thinks there is a causal relationship between these two variables.
Is he correct? You must give reasons for your answer. **(2)**

No, he is not correct.

Buying an umbrella does not cause a road accident. There is not a causal relationship between the two variables. Both variables are affected by the weather, and by rain in particular.

You need to consider if there is any reason why one should affect the other.

Think if there are other factors which may cause both variables to increase.

Now try this

tier F&H

1 Which of these pairs of variables are likely to have a causal relationship?

A Number of units of electricity used and outside temperature

B IQ and shoe size

C Number of tooth fillings and sugar consumption

D Weight of item and cost **(2)**

2 Greg created a scatter diagram to compare height and performance in an intelligence test. The scatter diagram showed positive correlation.

Discuss possible reasons why the scatter diagram could show positive correlation. **(2)**

Line of best fit

You can use a **line of best fit** to summarise the relationship shown on a scatter diagram. The line of best fit can be used to predict values.

Drawing a line of best fit

You need to be able to draw a line of best fit **by eye** and by drawing through the **mean point**.

This scatter diagram shows students' estimates of the lengths of some lines whose true lengths were known by their teacher.

The coordinates of the mean point $(\bar{x}, \bar{y})$ for a set of data values for x and y is given by:

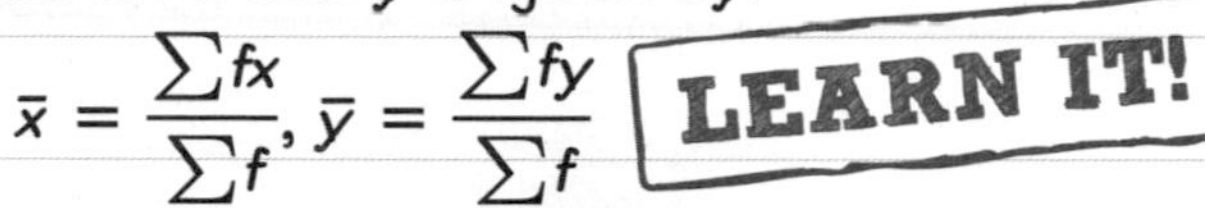

$$\bar{x} = \frac{\sum fx}{\sum f}, \bar{y} = \frac{\sum fy}{\sum f}$$

LEARN IT!

The line of best fit should extend before the first point and after the last point, so it covers all of the data.

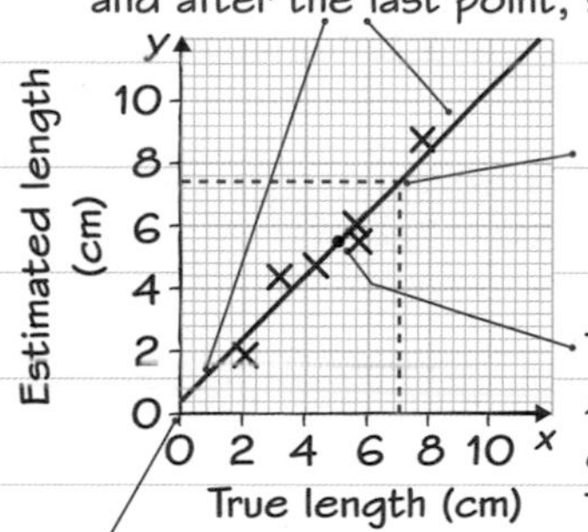

The line can be used to read off an estimate of one variable, given the value of the other. This line shows that a true length of 7 cm was, on average, estimated as 7.3 cm.

This point is the mean for both x and y. If these means are given in a question, draw the line through this point. The coordinates are given as $(\bar{x}, \bar{y})$.

The line of best fit does not need to pass through (0, 0).

Worked example

The scatter diagram gives some information about the prices (£y) of some cars and their ages (x years).

$\bar{x} = 4$, $\bar{y} = 4600$.

(a) Draw a line of best fit on the diagram. **(1)**

(b) Use the line of best fit to predict the price of a car that is $5\frac{1}{2}$ years old. **(1)**

Price = £3400

Read off the line of best fit at $x = 5.5$

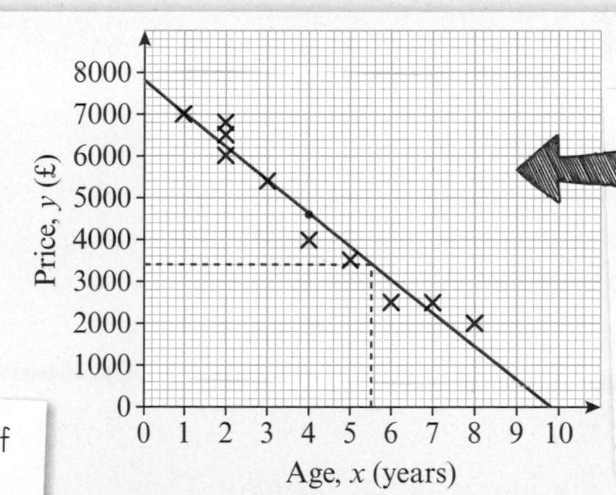

If you are given a table of data, you may need to work out the coordinates of the mean point.

Now try this

tier F&H

1 The scatter diagram shows information about arm length and leg length.

(a) Draw a line of best fit. **(1)**

Mo has a leg length of 65 cm.

(b) Estimate his arm length. **(1)**

Arm length (cm): 0, 10, 20, 30, 40, 50

Leg length (cm): 30, 40, 50, 60, 70, 80

You are not given a table of values or the mean values, so you must draw the line of best fit by eye.

2 The table shows the times students spent revising for an exam and the marks they achieved.

Student	A	B	C	D	E	F	G	H	I	J
Time	5	7	9	4	6	2	7	5	1	8
Mark (%)	80	75	92	60	70	40	84	68	35	95

(a) Draw a scatter diagram for the data. **(3)**

(b) Find the mean point and mark it on your diagram. **(2)**

(c) Draw a line of best fit and use it to predict the mark of a student who spends 3 hours revising. **(2)**

Interpolation and extrapolation

You can use a line of best fit to **estimate** values within the range of the given data. This is called **interpolation.**

You can also use it to **predict** values beyond the range of the given data. This is called **extrapolation.** Extrapolation is less reliable than interpolation.

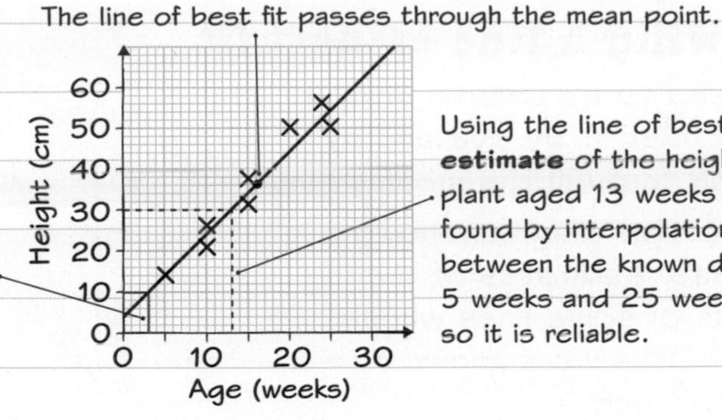

Worked example

Calculate the mean point of the data first. The line of best fit must go through the mean point.

The scatter diagram gives information about the heights and lifetimes of six candles.
One further candle has an initial height of 7 cm and a lifetime of 12 hours.

(a) Plot this point on the diagram. **(1)**

(b) Using the mean point, draw the line of best fit. **(2)**

For all seven candles the mean height is 9 cm and the mean lifetime is 15 hours.

A candle has an initial height of 15 cm.

(c) Estimate its lifetime. **(1)**

Using the line of best fit, the lifetime is 21.5 hours.

This is **interpolation**. It is reliable as it uses a value within the range of the measurements.

Another candle has an initial height of 22 cm.

(d) Predict its lifetime. **(1)**

Using the line of best fit, the lifetime is 29 hours.

This is **extrapolation**. It is **not** reliable as it uses a value outside the range of the measurements.

Lifetime (hours) / Initial candle height (cm)

Now try this

The diagram shows information about the prices paid for some second-hand cars and the ages of the cars. The line of best fit has been drawn.

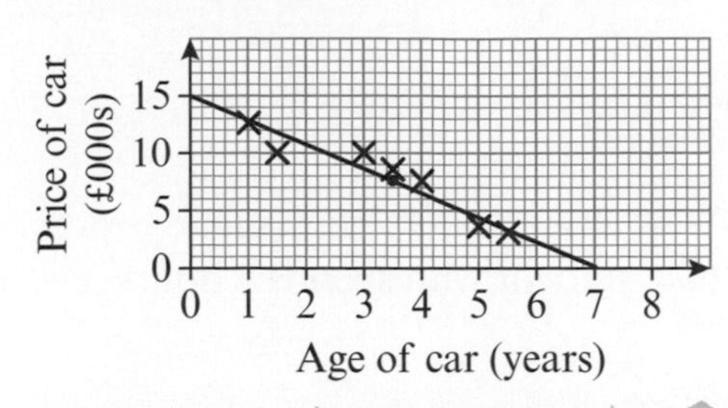

A second-hand car is sold when it is two years old.

(a) (i) Estimate its price.
(ii) Comment on the reliability of your estimate. **(2)**

Another second-hand car is sold when it is 6.5 years old.

(b) (i) Predict its price.
(ii) Comment on the reliability of your estimate. **(2)**

The equation of a line of best fit 1

You need to be able to work out the **equation of the line of best fit** in the form **$y = ax + b$**, where a is the **gradient** of the line and b is the **intercept** on the y-axis.

The scatter diagram gives information about the cost of a fruit cake and the percentage of fruit in the cake. The line of best fit has been drawn.

To find the gradient, choose two points on the line and draw a right-angled triangle. Then find the length of the height and base.

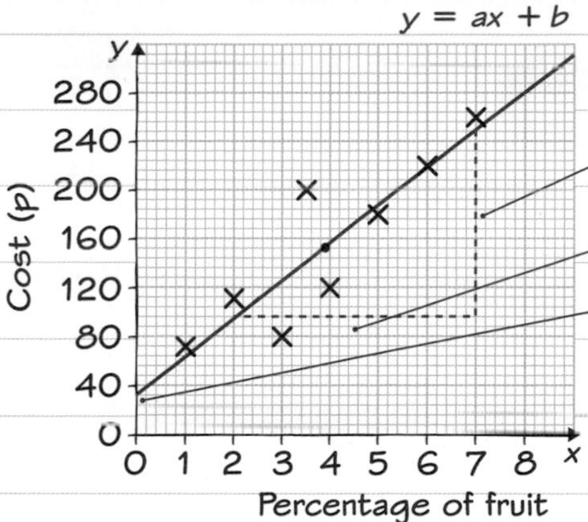

a = gradient = $\frac{\text{height}}{\text{base}}$

height = difference in y-values

base = difference in x-values

b = intercept on the y-axis

When finding the base and the height, use the scale. Counting squares may give the wrong answer.

The gradient is a measure of the **rate of increase** of the y-value with the x-value.

The y-intercept, b, is the value of y when the value of x is 0.

If the correlation is:

- positive, the value of a (the rate of increase) will be positive
- negative, the value of a will be negative.

Worked example (tier H)

The scatter diagram gives information about the temperature, x°C, and the number of sales of drinks, y, from a shop each day for 10 days.

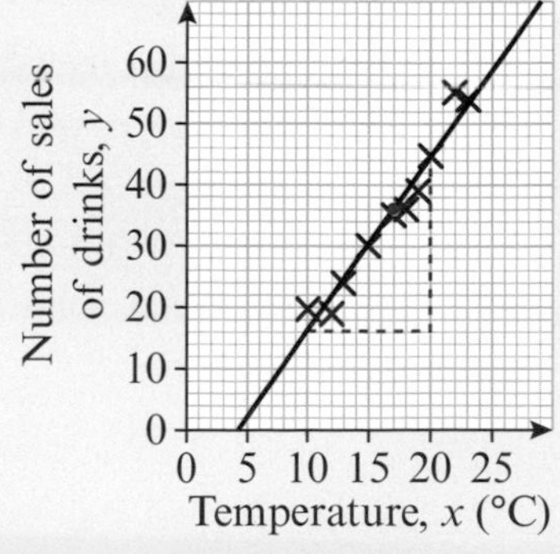

(a) Find the equation of the line of best fit. **(3)**

a = gradient = $\frac{\text{height}}{\text{base}} = \frac{45 - 16}{20 - 10} = 2.9$

$b = y - ax = 16 - 2.9 \times 10 = -13$

Equation of the line of best fit is

$y = 2.9x - 13$

(b) What information about the drink sales is given by the gradient of the line of best fit? **(1)**

With each temperature increase of 1°C, 2.9 more drinks were sold.

Think about what the gradient means in the context of the question.

Now try this (tier H)

The scatter diagram gives information about the sizes (S litres) of some car engine and the times (t seconds) it takes the cars to accelerate to 30 mph.

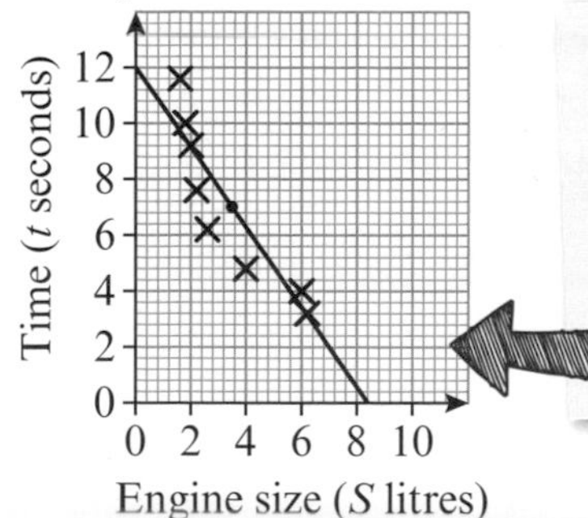

Remember, the correlation is negative so the gradient will also be negative.

A line of best fit has been drawn on the scatter diagram.

(a) Find the equation of the line of best fit. **(1)**

(b) What information is given by the gradient? **(1)**

(c) Why might the intercept in the equation not be realistic? **(1)**

Had a look ☐ Nearly there ☐ Nailed it! ☐

The equation of a line of best fit 2

The line of best fit is sometimes known as the **regression line.** You need to be able to draw a regression line on a scatter diagram when given the equation of the line.

Worked example

tier H

The scatter diagram gives information about the percentage marks of 12 students in a maths test and a science test. The equation of the regression line is $y = 1.05x - 3$

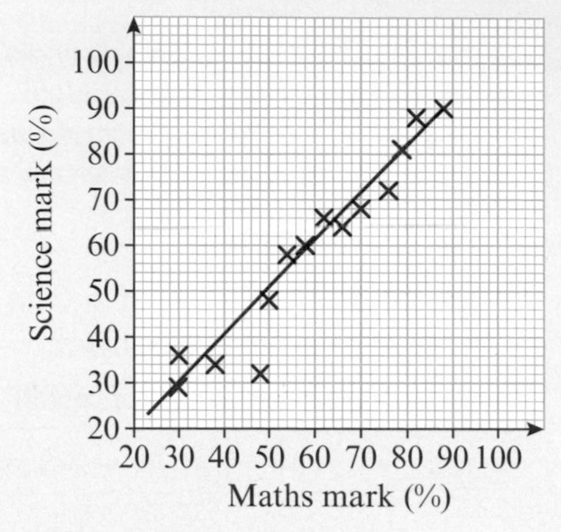

Find the coordinates of two points on the regression line using the equation of the line. Join the points with a straight line. Choose values of x that are easy to substitute into the equation and which are towards the lower and upper ends of the plotted values.

(a) Draw this line on the scatter diagram. **(2)**

When $x = 30$, $y = 1.05 \times 30 - 3 = 28.5$

When $x = 80$, $y = 1.05 \times 80 - 3 = 81$

Plot the point (30, 28.5)
Plot the point (80, 81)

(b) Interpret the y-intercept of the line. **(1)**

The regression line will cross the y-axis at -3, which implies that a student who scores 0% on the maths test would score -3% on the science test. This is not a possible score, which shows the regression line is not a reliable predictor below the range of given values.

Remember to use the value -3 from the equation of the regression line as the scatter diagram does not start from (0, 0).

(c) Interpret the value of the gradient of the line. **(1)**

For every extra 1% scored on the maths test, a student would be expected to score 1.05% more on the science test.

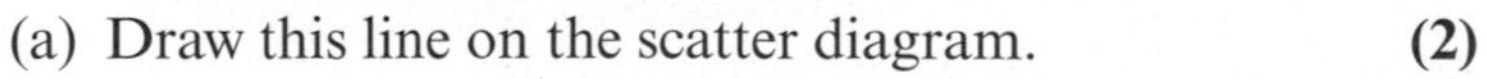

Now try this

tier H

The scatter diagram shows the age of a car, x and its value y. The equation of the regression line is $y = 20000 - 1500x$.

(a) Draw this line on the scatter diagram. **(2)**

(b) Interpret the y-intercept of the line. **(1)**

(c) Interpret the value of the gradient of the line. **(2)**

(d) Use the equation of the regression line to predict the value of the car after 15 years.

Comment on the reliability of your answer. **(2)**

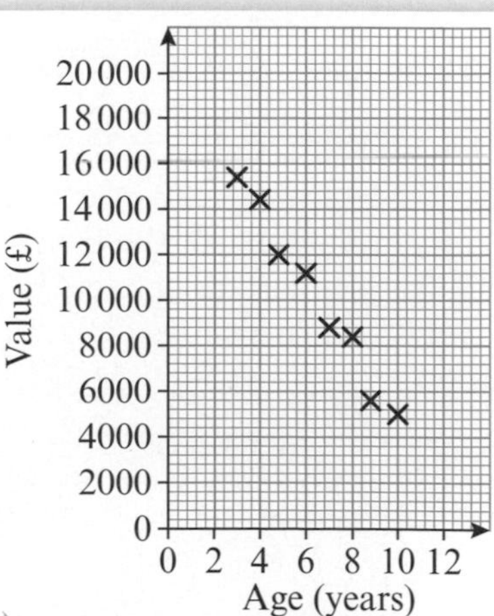

Spearman's rank correlation coefficient

Spearman's rank correlation coefficient, r_s, measures the strength of the correlation between two sets of data. You need to be able to interpret Spearman's rank correlation coefficient in the context of the problem.

The value for Spearman's rank correlation coefficient lies between −1 and +1. The further the value is from zero, the stronger the correlation.

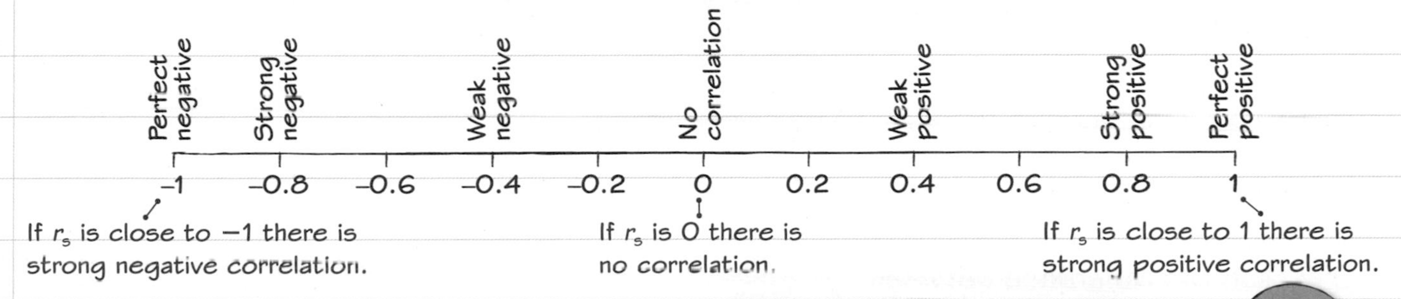

Scatter diagrams

A scatter diagram can give an indication of correlation, but the formula for calculating Spearman's rank correlation coefficient (see page 68) gives an accurate value.

The correlation is positive but is not perfectly positive so the most likely value is 0.8.

Worked example (tier F&H)

Which is the most likely Spearman's rank correlation coefficient for the data shown in this scatter diagram? **(1)**

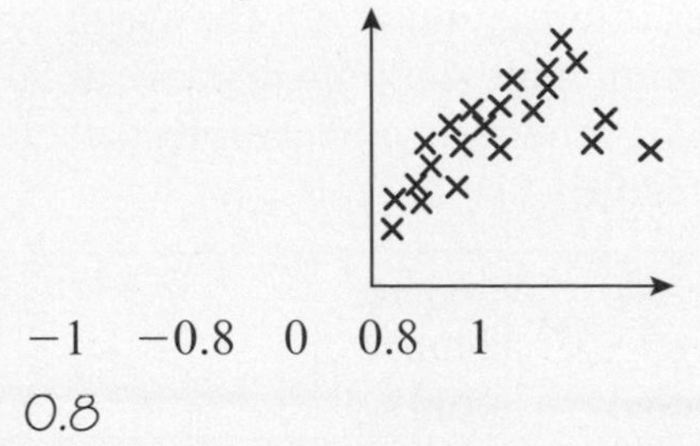

−1 −0.8 0 0.8 1

0.8

Worked example (tier F&H)

Two judges scored the dives of two divers in a diving competition.
Ed and Kate calculated the value of Spearman's rank correlation coefficient for the ranks given by the judges.
Ed got a value of 0.7 and Kate got a value of 1.2.

(a) Why is Kate's value incorrect? **(1)**

The value of Spearman's rank correlation coefficient always lies between −1 and +1 (inclusive).

(b) Ed's value is correct. Describe and interpret his value. **(2)**

The value shows moderately strong positive correlation, so the judges were in reasonably good agreement with their scores for the divers.

State whether the correlation is positive or negative and strong or weak.

Now try this (tier F&H)

10 dancers competed in a dancing competition. Two judges ranked the 10 dancers on their first dance. The Spearman's rank correlation coefficient value for the rankings was 0.75.

(a) Describe and interpret this value. **(2)**

The judges then ranked the 10 dancers on their second dance. The Spearman's rank correlation coefficient for the second dance was calculated as 0.68.

(b) Compare the two correlation coefficients. **(2)**

Explain what the correlation means in the context of the question.

Had a look ☐ Nearly there ☐ Nailed it! ☐

Calculating Spearman's rank correlation coefficient

You might need to **rank** the data values before you can calculate Spearman's rank correlation coefficient.

Ranking data

The table shows the position in the league and the number of games that each of ten teams lost.

To convert the number of games lost into a rank, add an extra row labelled Rank and write numbers in order of the number of games lost.

d is then the difference between each team's league position and your rank.

League position	1	2	3	4	5	6	7	8	9	10
Number of games lost	2	9	13	16	14	11	19	18	20	26
Rank	1	2	4	6	5	3	8	7	9	10
d	0	0	1	2	0	3	1	1	0	0

Then you can work out the rank correlation coefficient

using $r_s = 1 - \frac{6\sum d^2}{n(n^2 - 1)} = 1 - \frac{6 \times 16}{10(10^2 - 1)} = 0.90$

The value of 0.90 shows a very strong positive correlation between league position and number of games lost.

Worked example

tier H

The table gives information about the age of tyres on a car and the minimum stopping distance at 50 mph. Work out Spearman's rank correlation coefficient and interpret the result. **(5)**

In the exam you must remember to write down your working.

Car	Age of tyres (months)	Stopping distance (m)	d	d^2
A	10 (1)	42 (1)	1 − 1 = 0	0
B	11 (2)	49 (3)	2 − 3 = −1	1
C	15 (3)	47 (2)	3 − 2 = 1	1
D	20 (4)	53 (6)	4 − 6 = −2	4
E	24 (5)	51 (5)	5 − 5 = 0	0
F	28 (6)	58 (7)	6 − 7 = −1	1
G	30 (7)	50 (4)	7 − 4 = 3	9
H	32 (8)	60 (8)	8 − 8 = 0	0
				16

You can write the ranks beside the values in the table. Sometimes there are separate columns for you to put them in.

You can add an extra row to add up your values for d^2.

Spearman's rank correlation coefficient

$= 1 - \frac{6 \times 16}{8 \times (8^2 - 1)} = 0.81$

Spearman's rank correlation coefficient for the age of tyres and stopping distance is 0.81. There is strong positive correlation between the age of tyres and the stopping distance.

Now try this

tier H

The table gives information about the length and the weight of eight large birds.

Length (cm)	28	40	48	60	63	78	98	120
Weight (kg)	0.4	0.7	0.6	1.1	1.0	2.2	4.1	5.6

(a) Calculate Spearman's rank correlation coefficient. **(3)**

(b) Interpret your answer to part (a). **(2)**

Pearson's product moment correlation coefficient

Pearson's product moment correlation coefficient (PMCC), r, is a measure of linear correlation which can be used to measure the strength of the association between sets of data. You need to be able to interpret the PMCC in the context of the data.

The value of r always lies between −1 and +1 and tells you how far the data points are from the linear regression line. The further it is from 0, the stronger the correlation.

- If r is close to +1 there is strong positive correlation.
- If r is close to 0 there is no correlation.
- If r is close to −1 there is strong negative correlation.

PMCC is similar to Spearman's rank correlation coefficient. You need to understand the differences and similarities between them. You don't need to calculate the PMCC.

Differences

- PMCC tests for **linear** correlation.
- Spearman's rank correlation coefficient tests for **any** correlation, including points lying on the same **curve**.

Similarities

- Both measures lie between −1 and +1.
- The further the value is from 0, the stronger the correlation.

In these scatter diagrams, Spearman's rank correlation coefficient shows perfect correlation because the x-values and y-values have the same rank. The PMCC varies according to how closely the values fit a **straight** line.

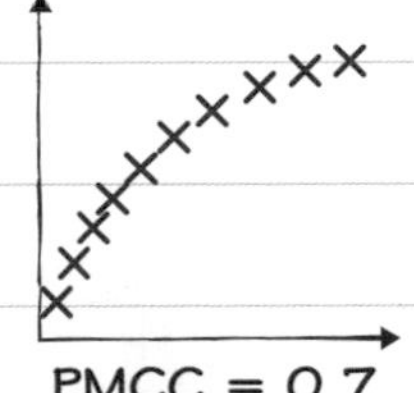
PMCC = 0.7
Spearman's rank = 1

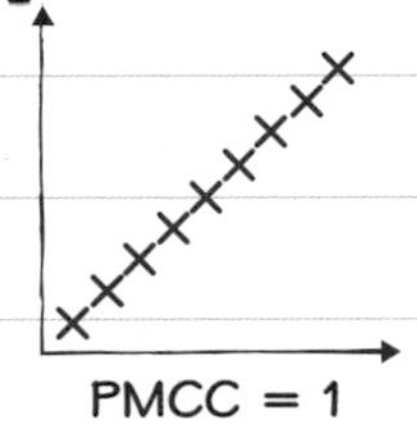
PMCC = 1
Spearman's rank = 1

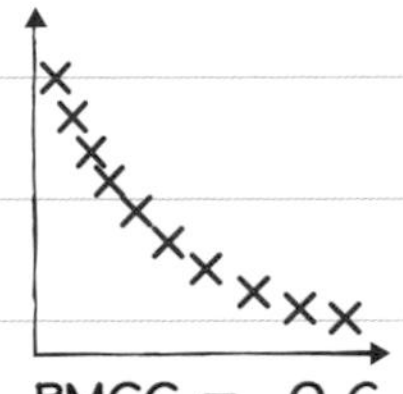
PMCC = −0.6
Spearman's rank = −1

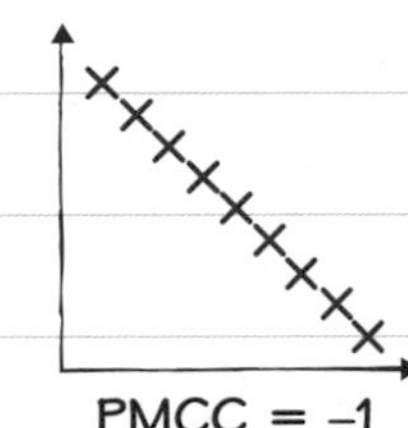
PMCC = −1
Spearman's rank = −1

Worked example

tier H

Statistical software calculated Spearman's rank correlation coefficient and Pearson's product moment correlation coefficient for the data shown in this scatter diagram.

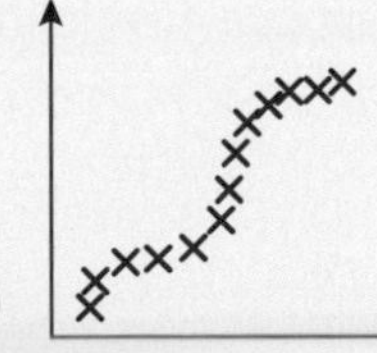

(a) Identify one value in each row to show the most likely **pair** of correlation coefficients for this data. **(2)**

Spearman's rank correlation coefficient	−1	−0.8	0	0.8	(1)
Pearson's product moment correlation coefficient	−1	−0.8	0	(0.8)	1

Spearman's rank correlation coefficient = 1
Pearson's product moment correlation coefficient = 0.8

(b) Explain your choice of answers. **(1)**

Both correlation coefficients are positive. PMCC is less strong because it measures closeness to a linear model. Spearman's rank correlation coefficient is 1 because the rank orders for the x-values and the y-values are the same.

Now try this

Describe scatter diagrams with these Pearson product moment correlation coefficients.

(a) −1 **(1)** (b) 0.25 **(1)** (c) −0.75 **(1)**

Had a look ☐ Nearly there ☐ Nailed it! ☐

Line graphs and time series

Time series graphs are **line graphs** which show variation over time. Time is plotted on the horizontal axis.

This time series graph shows how car sales varied in the first and second halves of the year from 2013 to the first half of 2017.

Time series graphs will often show variations over the four **quarters** of a year. Each quarter is a period of three months. Quarters are usually shown as Q1, Q2, Q3 and Q4.

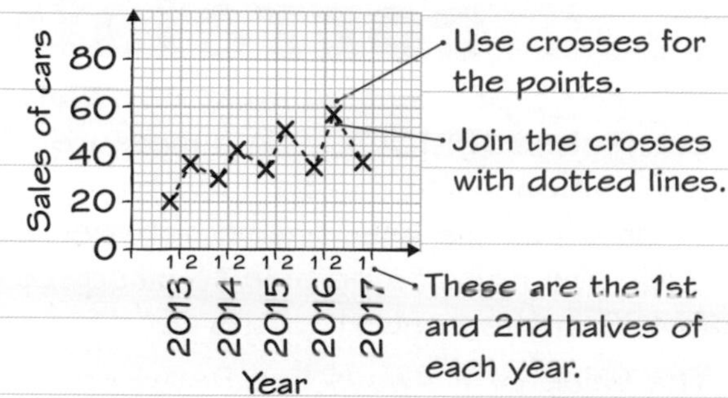

For every year, the number of sales was higher in the second half of the year.

Worked example

tier F

The graph shows information about the number of moths caught each quarter in 2015 and 2016. The table gives similar information for 2017.

	2017			
Quarter	1	2	3	4
Number caught	17	24	36	14

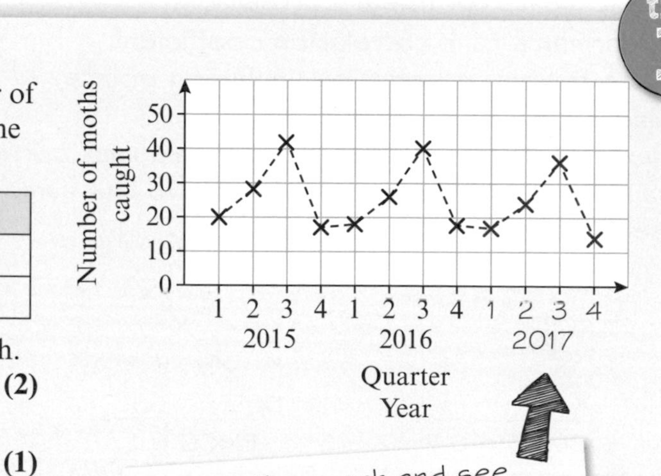

(a) Add this information to the time series graph. **(2)**

Look at the graph and see how any regular pattern varies.

(b) When was the greatest number of moths caught? **(1)**

In Quarter 3 of 2015.

(c) Give a possible reason why fewer moths were caught in Q1 and Q4 each year. **(1)**

These quarters contain the coldest months of the year and some moths hibernate in winter.

Always answer in context.

Now try this

tier F

The table gives information about the number of wild flowers in a park.

	2016				2017			
Quarter	1	2	3	4	1	2	3	4
Number	14	25	34	23	8	23	37	20

(a) Plot this information on a time series graph. **(2)**

(b) Write down the quarter with the highest number of wild flowers. **(1)**

(c) Comment on the graph. **(2)**

Trend lines

Times series graphs are useful for studying the **trend** of the data. A **trend line** shows the general trend of the data over time.

You need to be able to draw and use a trend line on a time series graph to interpret:

- **upward** (rising or increasing) trend
- **downward** (falling or decreasing) trend
- **level** (or no increase or decrease) trend.

Rising trend

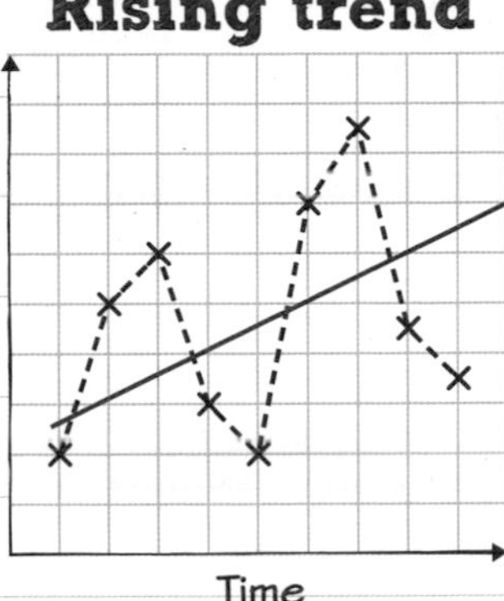

Falling trend

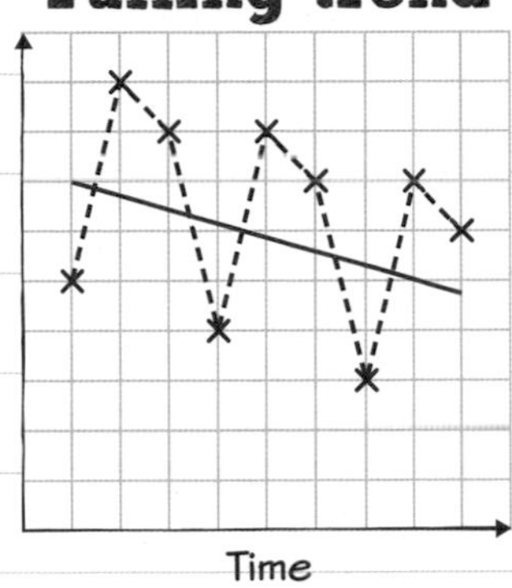

Level trend

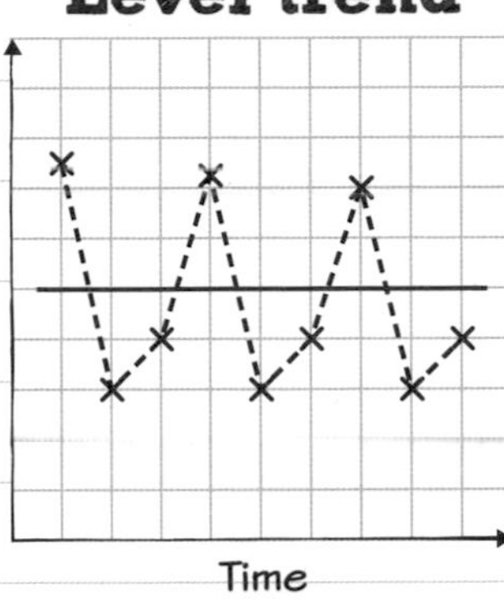

Worked example

tier F

Describe and interpret the trend in the number of car sales from 2013 to 2017. **(2)**

The trend line is upwards (rising).

The number of car sales is increasing with time.

Your answer must interpret the trend in the context of the question.

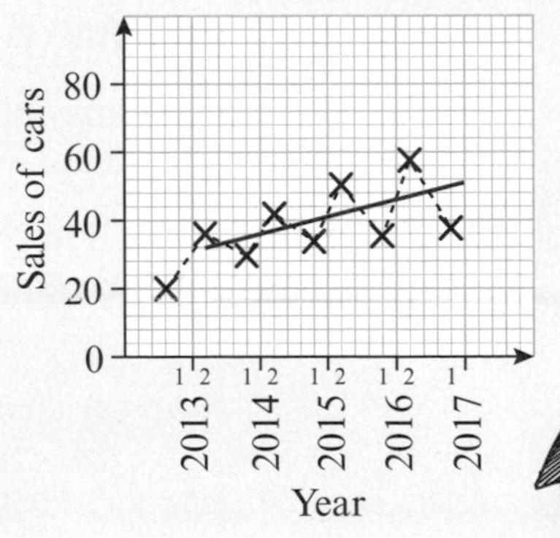

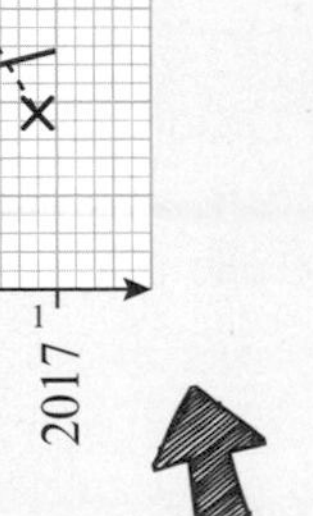

You can draw a **trend line** by eye. Place the line roughly halfway between the highest and lowest points for each year.

Now try this

A headteacher recorded the number of student absences each day for the first three weeks of a school term.

Week	1					2					3				
Day	M	Tu	W	Th	F	M	Tu	W	Th	F	M	Tu	W	Th	F
Absences	9	8	7	6	15	14	11	10	9	20	18	14	14	12	24

(a) Draw a time series graph for this data. **(3)**

(b) Draw a trend line on the graph. **(1)**

The headteacher is worried about the number of absences on Fridays.

(c) Use the time series graph to comment on whether the headteacher is right to be worried. **(2)**

 Had a look ☐ Nearly there ☐ Nailed it! ☐

Variations in a time series

A time series graph may show **variations** in its pattern. These may be

- a **general trend** (as shown by the trend line)
- **seasonal variations** (a pattern that repeats across any regular period).

Identifying variations

Variations are values which are either above or below a trend line.

A trend line has been added to the time series graph you met on page 70, for the number of moths caught from 2015 to 2017.

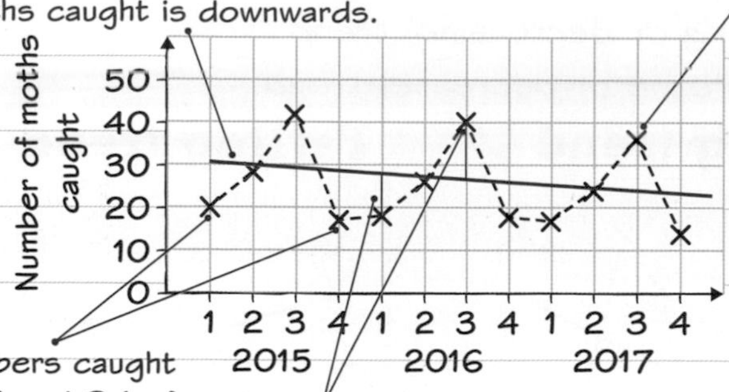

Worked example

tier F&H

The table shows the quarterly gas bills for a house with gas central heating and a gas cooker over a three-year period.

Year	1				2				3			
Quarter	1	2	3	4	1	2	3	4	1	2	3	4
Bill (£)/hrs	290	140	45	180	370	180	55	205	420	210	70	245

(a) Draw a time series graph of this data. **(3)**

(b) Draw a trend line on the graph. **(1)**

(c) Describe the variations shown in the graph. Suggest a reason why these seasonal variations take place. **(2)**

The graph shows an increasing trend – the cost of gas has increased each year. There is a seasonal variation with the gas bills being higher than the trend value in the first quarter and lower in the second, third and fourth quarters. More gas is used for heating in the colder months.

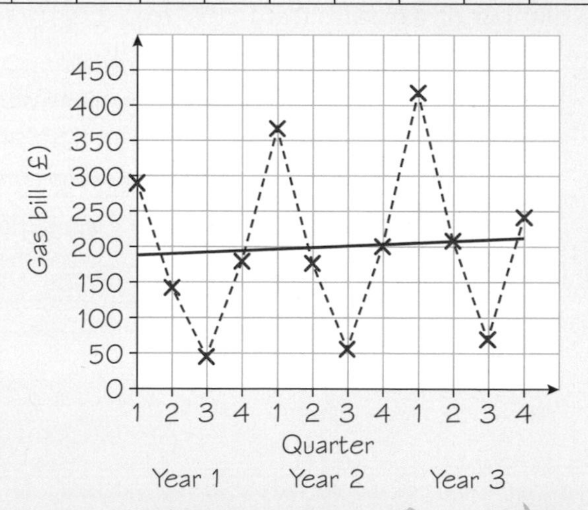

Now try this

tier F&H

The line graph shows the number of tickets sold each day for two weeks in a small local cinema. Comment on the trend and seasonal variations. **(2)**

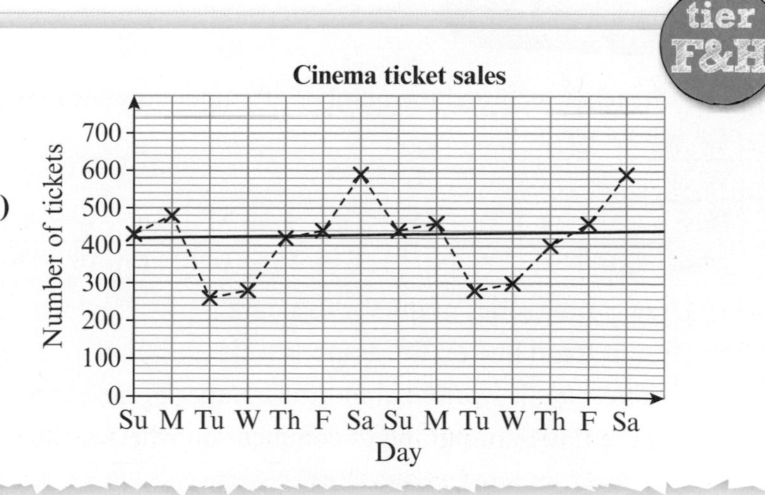

Moving averages 1

Time series may show large variations with upward and downward peaks. This can make it difficult to see the trend of the data and to draw a trend line.

A **moving average** is a way of smoothing out the data in a time series. It makes it easier to account for seasonal variation and to see what the trend is.

Calculating the moving average

The table shows information about the numbers of shoes sold in a shop.

1. Calculate the first 4-point moving average. This is the mean of the first 4 consecutive values:

$$\frac{250 + 306 + 282 + 244}{4} = 270.5$$

2. Calculate the 2nd 4-point moving average using the 2nd, 3rd, 4th and 5th values.
3. Calculate the 3rd 4-point moving average using the 3rd, 4th, 5th and 6th values.
4. Write all these in the 4-point moving average column.

The moving averages do not vary as much as the original figures.

Year	Quarter	Number of shoes	4-point moving average
2016	1	250	
	2	306	
			270.5
	3	282	
			280.5
	4	244	
			284
2017	1	290	
	2	320	

The number of points in each moving average should cover one complete cycle of seasons.

Worked example

tier F&H

The table gives information about the numbers of fish caught from a lake.

Year	Quarter	Number caught (1000s)	4-point moving averages (1000s)
2016	1	20	
	2	28	
			26.75
	3	42	
			26.25
	4	17	
			25.75
2017	1	18	
			25.25
	2	26	
			25.50
	3	40	
	4	18	

(a) Complete the 4-point moving average column. **(2)**

(b) Describe the trend in the numbers caught. **(1)**

Looking at the moving average column, the number of fish caught is decreasing.

Find the first 4-point moving average by finding the mean of the first four values:

$$\frac{20 + 28 + 42 + 17}{4} = 26.75$$

Make sure you put the moving averages in the correct cells – the mid-point of the values they cover. The first 4-point moving average is placed **between** the 2nd and 3rd values to show that it is calculated from the first four values.

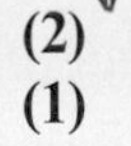

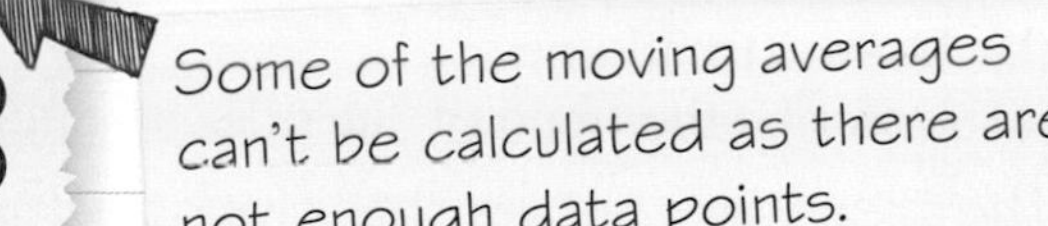

Now try this

tier F&H

The table gives information about the numbers of cars sold by season in 2016 and 2017.

	2016				2017			
Season	Spring	Summer	Autumn	Winter	Spring	Summer	Autumn	Winter
Number	35	20	53	42	43	33	57	38

(a) Work out the 4-point moving averages for this information. **(2)**

(b) Comment on any trend in the sales. **(1)**

Moving averages 2

You can draw a **trend line** by plotting the moving averages on a time series graph. The moving averages are plotted at the midpoint of the time intervals they cover.

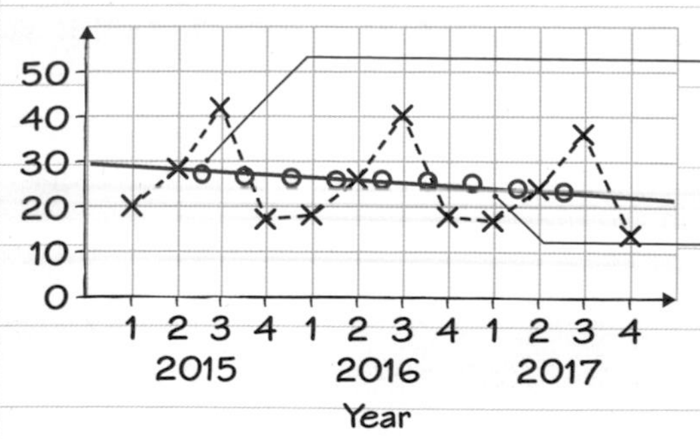

The 4-point moving averages are plotted halfway between the 2nd and 3rd points. So the 1st moving average (for quarters 1, 2, 3 and 4) is plotted at position 2.5. The next is at position 3.5.

To plot the trend line, draw the line of best fit through the moving average points by eye. The trend line should have roughly equal numbers of points above and below it. You do not join up the points for the moving averages.

A moving average is normally calculated over four points, but you can use other intervals.

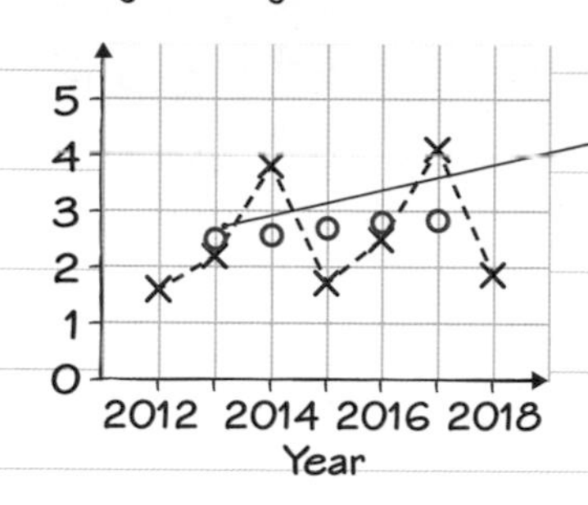

Each 3-point moving average is plotted against the middle point. So the average for 2012, 2013 and 2014 is plotted at 2013.

Worked example

tier H

This graph shows the sales of smoothies from a café.

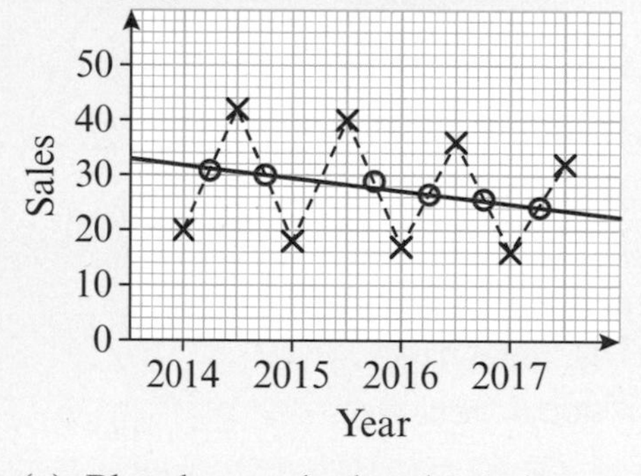

The 2-point moving averages are plotted halfway between the crosses.

(a) Plot these calculated 2-point moving averages on the time series. **(2)**

31	30	29	28.5	26.5	26	24

(b) Draw a trend line. **(1)**

(c) Describe the trend of the sales of smoothies over the years 2014 to 2017. **(1)**

The trend line shows that the sales are decreasing.

Answer the question in context here by referring to 'sales'.

Now try this

tier H

	2015			2016			2017	
Rainfall (cm)	102	156	142	106	157	135	110	169
3-point moving average		133	135	135	133	134	138	

(a) Plot the time series. **(2)**

(b) Plot the moving averages. **(2)**

(c) Draw the trend line. **(1)**

(d) Describe the trend. **(1)**

Seasonal variations

LEARN IT!

- **Seasonal variation** at a point = actual value − trend value.
- **Estimated mean seasonal variation** = mean of all the seasonal variations for that season.
- **Predicted value** = trend line value + estimated mean seasonal variation.

Worked example

tier H

The table and the graph show the sales of football boots in a sports shop for 2016 to 2018.

Year	Four-month period	Pairs of football boots	3-point moving average
2016	Jan-Apr	144	
	May-Aug	53	126
	Sept-Dec	182	120
2017	Jan-Apr	124	115
	May-Aug	39	112
	Sept-Dec	172	113
2018	Jan-Apr	128	113
	May-Aug	40	109
	Sept-Dec	158	

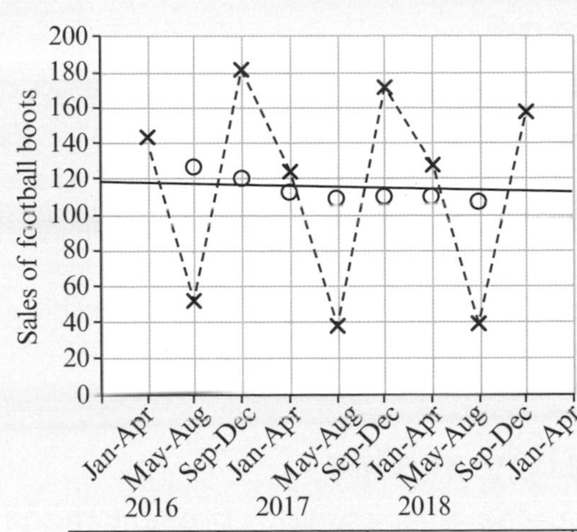

(a) Use the trend line and the mean seasonal variation to predict the number of pairs of boots in the period Jan-Apr 2019. **(4)**

Sammy wants to use this information to predict the sales of football boots in 2020.

(b) Discuss whether or not it would be appropriate to do so. **(2)**

(a)

Jan-Apr	Actual sales	Trend (from line)	Seasonal variation
2016	144	118	26
2017	124	116	8
2018	128	114	14

Mean seasonal variation for Jan-Apr

$= \frac{26 + 8 + 14}{3} = 16$

Predicted value for Jan-Apr 2019

$= 112 + 16 = 128$

(b) Not appropriate as the trend may not continue so far into the future.

Now try this

The table gives information about the number of students who enrol for a course in each term in 2015, in 2016 and in 2017.

Year	Term	Number of students	3-point moving average
2015	Autumn	320	
	Spring	170	190
	Summer	80	210
2016	Autumn	380	227
	Spring	220	253
	Summer	160	277
2017	Autumn	450	287
	Spring	250	307
	Summer	220	

(a) Draw the time series graph and plot the moving averages. **(4)**

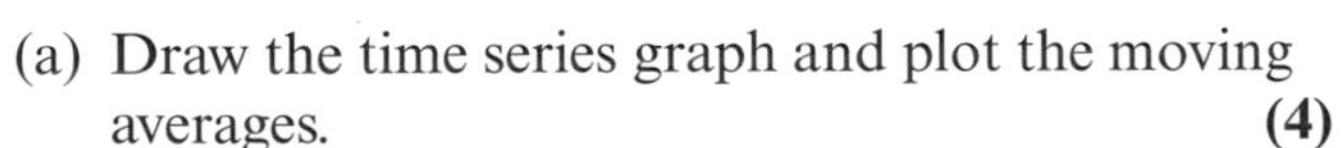

(b) On the time-series graph, draw a trend line for the 3-point moving averages. **(1)**

(c) (i) Use your trend line to find an estimate for the mean seasonal variation in numbers enrolling for the Autumn Term. **(2)**

(ii) Predict the number of students who will enrol in the Autumn Term of 2018. **(2)**

The meaning of probability 1

Probability is concerned with the chances of **outcomes** and **events.**

- An **outcome** is the result of a trial (such as rolling a dice once, for which the outcome is the number rolled).
- An **event** is a particular result which may contain one or more outcomes (such as getting a number more than 4 when you roll a dice).

Likelihood

You need to be able to use these words to describe the likelihood of an event happening.

Impossible	**Unlikely**	**Evens**	**Likely**	**Certain**
(can't happen)		(as likely to happen as not)		(must happen)

Increasing likelihood →

You can also use 'very unlikely' for something that is nearly impossible and 'very likely' for something that is almost certain.

Probability scales

You can also use numbers to represent the probability that an event will occur.

0 is impossible, 0.5 is evens and 1 is certain.

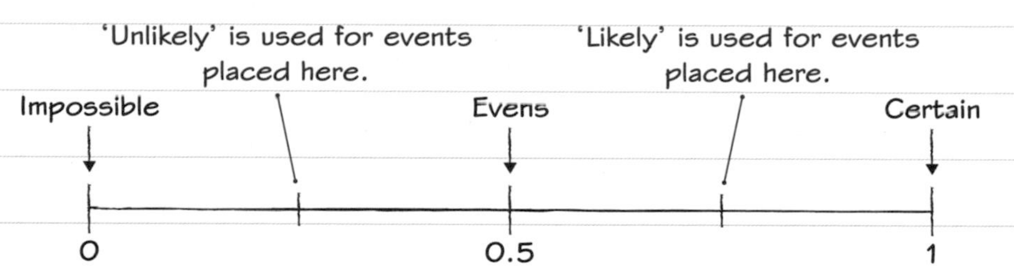

Worked example

tier F

Use a suitable word to describe the likelihood of each of these events.

(a) It will snow in London in May. **(1)**

Unlikely

(b) The next person born in the UK will be a boy. **(1)**

Evens

(c) You roll an ordinary dice and get the number 7. **(1)**

Impossible

Sometimes you will be given a list of words to choose from and then you must choose one from the list.

Numbers on an ordinary dice are 1 to 6.

Now try this

tier F

On the probability scale, mark the probability of each event with a cross (×).

(a) You will have something to eat tomorrow. Label this cross **A**. **(1)**

(b) A teacher selected at random was born on a Monday. Label this cross **B**. **(1)**

(c) A fair 6-sided dice will show an even number when rolled. Label this cross **C**. **(1)**

0 ——— 0.5 ——— 1

The meaning of probability 2

Expressing probability

Probabilities can be written as fractions, decimals or percentages.

If the probability it will rain today is 60%, this can be written as

$P(R) = 60\%$

$= \frac{60}{100}$

$= 0.6$

P represents probability.

R represents rain. You can choose any suitable letter.

Expected frequency

If you know the probability of an event, you can predict how many times you would expect the event to happen in a given number of trials.

Expected frequency of event A

= P(A) × number of trials

LEARN IT!

The expected frequency is the number of times you expect the event to happen. It is not necessarily what will actually happen.

Probability for equally likely events

On a **fair** six-sided dice, each number from 1 to 6 is **equally likely** to be rolled.

In a full pack of 52 cards, all cards have equal probability of being picked at random.

If all possible events are equally likely:

LEARN IT!

The **probability** of an event

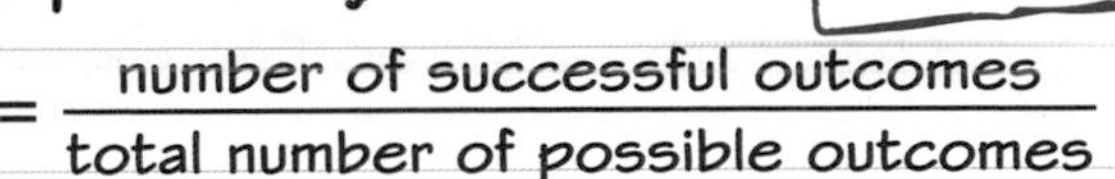

$$= \frac{\text{number of successful outcomes}}{\text{total number of possible outcomes}}$$

This bag contains three red beads and four black beads. One bead is taken out of the bag at random.

The probability it will be a red bead is found by counting the number of outcomes that are red.

B B B B R R R

3 of the 7 outcomes are R, so the probability is $\frac{3}{7}$

Worked example

tier F

A fair six-sided dice is rolled.

(a) Work out the probability of rolling a number greater than 4. **(1)**

P(number > 4) = $\frac{2}{6}$

There are six possible outcomes: 1, 2, 3, 4, 5, 6. Two of these are greater than 4 (5 and 6).

You do not need to write the fraction in its simplest form.

The dice is rolled 100 times.

(b) Work out an estimate for the number of times you would expect to roll a number greater than 4. **(2)**

Expected frequency = $\frac{2}{6} \times 100 = 33.3\ldots$

You would expect a number greater than 4 to be rolled 33 times.

Round your answer to an integer.

Now try this

tier F

The diagram shows a fair spinner. The spinner is spun exactly 60 times.

Work out an estimate for the number of times the spinner will point to blue. **(3)**

The meaning of probability 3

You need to be able to calculate probabilities from data given in **two-way tables.**

The table shows the sandwich orders at a deli one lunchtime. Each person chose one sandwich.

		Bread		
		White	Brown	Total
Filling	Turkey	4	5	9
	Jam	6	4	10
	Cheese	7	6	13
	Total	17	15	32

10 people chose jam. So if a person is picked at random, the probability that they chose jam is

$$P(\text{jam}) = \frac{\text{number of successful outcomes}}{\text{total number of possible outcomes}} = \frac{10}{32} = 0.3125$$

This is the total number of people. So if one person is picked at random, this is the total number of possible outcomes.

Worked example

tier F

This table gives information about the services done by a garage at the start of the year.

		Month			
		Jan	Feb	Mar	Total
Services	Tyres	15	8	4	27
	Exhaust	19	22	7	48
	Total	34	30	11	75

One of these services is selected at random for a quality check.

(a) Which service, tyres or exhaust, is more likely to be selected? **(1)**

Exhaust

(b) Write down the probability that the service was done in February. **(1)**

$\frac{30}{75} = \frac{2}{5}$

(c) Write down the probability that the service was tyres and it was done in January. **(1)**

$\frac{15}{75} = \frac{1}{5}$

The most likely service is the one with the higher total. There were 48 exhaust services and 27 tyre services. An exhaust service is more likely to be selected.

To find the probability that the service was done in February use:

$$P(\text{Feb}) = \frac{\text{number of successful outcomes}}{\text{total number of possible outcomes}} = \frac{30}{75} \left(= \frac{2}{5}\right)$$

There were 15 tyre services in January (number of successful outcomes) and 75 services done in total (total number of possible outcomes). The probability is $\frac{15}{75}$.

Now try this

tier F

This table gives information about some students' eating habits.

		Year group			
		Year 9	Year 10	Year 11	Total
Diet	Vegetarian	15	8	4	
	Not vegetarian	19	22	7	48
	Total			11	75

(a) Complete the table. **(2)**

One student is selected at random for a further interview.

(b) (i) Write down the diet that this student is most likely to have. **(1)**

(ii) Write down the probability that a student from Year 10 will be selected. **(1)**

(iii) Write down the probability that a vegetarian student from Year 11 will be selected. **(1)**

Experimental probability

You can find **estimates** of a probability by repeating an **experiment** many times.

Each experiment (or response to a survey) is a **trial**.

LEARN IT!

Estimated probability

$$= \frac{\text{number of trials with successful outcome}}{\text{total number of trials}}$$

The more times the experiment is repeated the more accurate the estimate can be.

The table shows the results of an experiment to discover whether a coin is biased.

Number of heads	34
Number of throws	50
Estimate	0.68

The results give an experimental probability of $\frac{34}{50} = 0.68$ for getting a head.

Experimental vs predicted

You can compare predicted results with results from an experiment to test for **bias**.

This table shows the predicted and experimental results of spinning a 4-sided spinner 100 times.

	1	2	3	4
Predicted	25	25	25	25
Experimental	21	22	33	24

The experimental figures are higher than expected for the number 3.
The spinner could be biased towards 3.

Worked example

tier F

Gary has a 4-sided spinner.
The sides of the spinner are labelled A, B, C and D.
The spinner is spun 200 times. The table shows the numbers of times the spinner landed on each side.

Side	A	B	C	D
Frequency	52	15	54	79

(a) If the spinner is fair how many times would Gary expect the spinner to land on side D? **(2)**

Expected number of times to land on D $= \frac{200}{4} = 50$

(b) Compare the results from the table with the expected results. **(2)**

The spinner lands on D more often than expected, on sides A and C about as many times as expected but a much lower number of times on side B. The spinner seems to be biased in favour of side D and against side B.

Now try this

tier F

Abel, Beth and Connor each test the same dice for bias. Their results are given in the table.

	Abel	Beth	Connor
Number of 2s	15	24	36
Number of rolls	60	90	150
Estimate of probability of a 2			

Remember: the more data, the more accuracy.

Connor claims that his results will give the most accurate estimate.

(a) Explain why Connor is correct. **(1)**

(b) Complete the last row of the table. **(2)**

(c) (i) Explain how it is possible from these results to get a better estimate than Connor's.
(ii) Find the value of the best estimate. **(2)**

Using probability to assess risk

Risk and accidents

The **risk** of a particular type of accident or problem with a machine is obtained from data. The risk of an accident is its experimental probability.

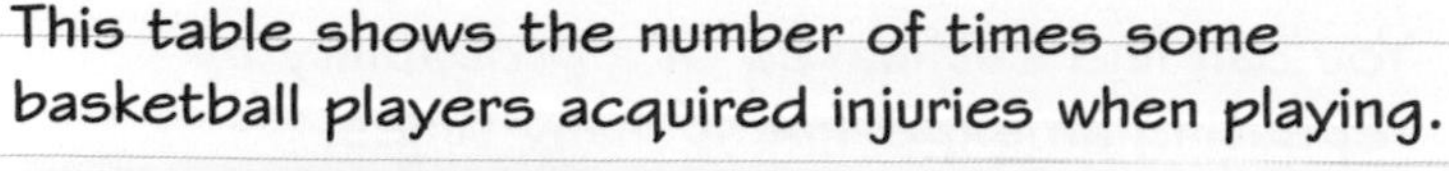

This table shows the number of times some basketball players acquired injuries when playing.

Injury	Fingers	Feet	Knees
Number of injuries	6	8	3
Games played	50	46	50

Write the risks as decimals. It makes them easier to compare.

The risk of an injury to fingers is $\frac{6}{50} = 0.12$

The other two injury types have risks of 0.17 and 0.06

$$\text{Risk of event} = \frac{\text{number of trials in which event happens}}{\text{total number of trials}}$$

LEARN IT!

You can compare the risk of an event happening for different groups.

- The **absolute risk** is the probability of an event happening.
- The **relative risk** is how many times more likely an event is to happen for one group compared to another group.

Relative risk for a group

$$= \frac{\text{risk for those in the group}}{\text{risk for those not in the group}}$$

Divide the risk for patients on the old treatment by the risk for patients on the new treatment.

Worked example

tier F&H

In a study of two cancer treatments on patients, the new treatment has a probability of 15% of being ineffective. The old treatment has a probability of 35% of being ineffective. What is the relative risk of the treatment being ineffective for patients using the old treatment compared to the new treatment? **(3)**

Relative risk on old treatment

$= \frac{0.35}{0.15} = \frac{7}{3} = 2.33...$

The relative risk is $2\frac{1}{3}$, so the risk of the treatment being ineffective is $2\frac{1}{3}$ times as high for patients on the old treatment.

Now try this

tier F&H

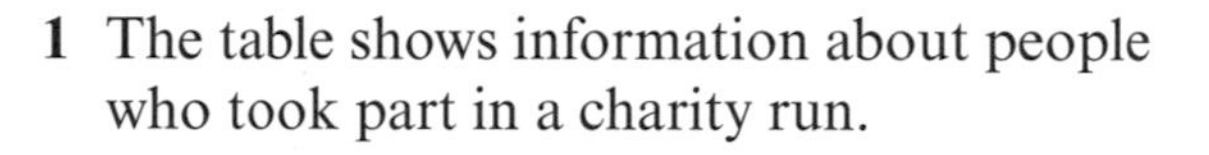

1 The table shows information about people who took part in a charity run.

	Warmed up before run	Did not warm up before run
Pulled a muscle	12	45
Did not pull a muscle	84	72

Use the information to find:

(a) the relative risk of pulling a muscle if you did not warm up before the run compared with if you did warm up before the run **(4)**

(b) the absolute risk of pulling a muscle in the charity run. **(2)**

2 A website for ambulance response times in the UK shows these results for two different postcodes in Manchester for 2014.

Manchester postcode	Percentage of life-threatening calls responded to in 8 minutes or less
M28	63%
M11	86%

Source: ambulanceresponsetimes.co.uk

(a) Work out the relative risk that an ambulance takes more than 8 minutes to respond in M28 compared with M11. **(3)**

Tia says, 'The relative risk shows that the risk of an ambulance taking over 8 minutes to respond in M28 is over 2 times greater than the risk of it taking over 8 minutes to respond in M11.'

(b) Use your answer to part (a) to explain whether or not Tia's conclusion is correct. **(1)**

Sample space diagrams

A **sample space**, S, is a list of all the possible outcomes of a trial.

To use a sample space, the outcomes must be **equally likely**.

If a fair dice is rolled once, the sample space is the list of numbers 1, 2, 3, 4, 5, 6.

If the event is 'any even number' then the possible members of the sample space that are included in the event are 2, 4 and 6.

You can use a table to represent the sample space for two events. This is often called a **sample space diagram**.

If two fair dice are each rolled once, the sample space consists of all the possible pairs:

(1, 1) (1, 2) (1, 3) (1, 4) (1, 5) (1, 6)
(2, 1) (2, 2) (2, 3) (2, 4) (2, 5) (2, 6)
(3, 1) (3, 2) (3, 3) (3, 4) (3, 5) (3, 6)
(4, 1) (4, 2) (4, 3) (4, 4) (4, 5) (4, 6)
(5, 1) (5, 2) (5, 3) (5, 4) (5, 5) (5, 6)
(6, 1) (6, 2) (6, 3) (6, 4) (6, 5) (6, 6)

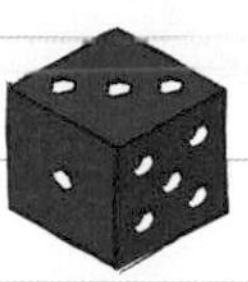

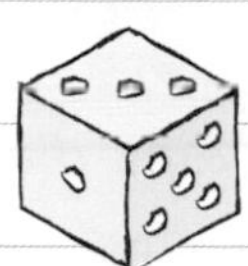

If the event is 'any double' then the successful outcomes from the sample space are (1, 1), (2, 2), (3, 3), (4, 4), (5, 5) and (6, 6).

If the event is 'a total of 4' then the successful outcomes are (1, 3), (2, 2) and (3, 1). You have to include (1, 3) **and** (3, 1) to describe the different orders of results.

Worked example

tier F

The diagram shows two spinners. Spinner A is spun once.

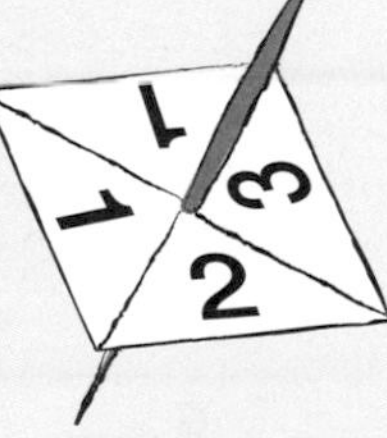

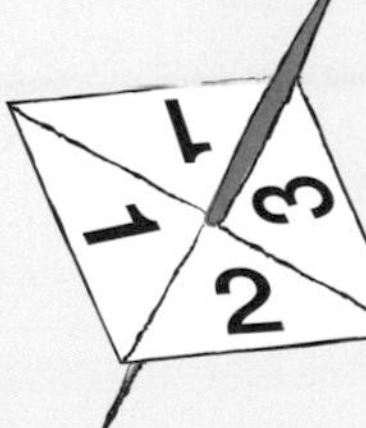

Spinner A Spinner B

(a) Write down a suitable sample space. **(1)**

1, 1, 2, 3

Spinner A and spinner B are each spun once.

(b) Write down a suitable sample space. **(2)**

		Spinner A			
		1	1	2	3
Spinner B	1	(1, 1)	(1, 1)	(1, 2)	(1, 3)
	1	(1, 1)	(1, 1)	(1, 2)	(1, 3)
	2	(2, 1)	(2, 1)	(2, 2)	(2, 3)
	3	(3, 1)	(3, 1)	(3, 2)	(3, 3)

Arrange the pairs in a table to make sure you don't miss any successful outcomes.

(c) List all the pairs in the sample space which have a total of 3. **(1)**

There are four pairs: (1, 2), (1, 2), (2, 1) and (2, 1).

There are some repeats in this list. This is not an error: they **should** be included.

Now try this

tier F

1 A 1p coin and a 2p coin are each spun once.

(a) Write out the sample space. **(1)**

(b) Which one of 2 heads, 2 tails, or 1 head and 1 tail is the most likely? **(1)**

2 There are three people in a room: Anna, Bruce and Connor. Two people are picked from the three.

(a) Write down all six elements of the sample space. **(1)**

(b) Which is more likely – two men or a man and a woman? **(1)**

Venn diagrams

A **Venn diagram** is a way of showing how items are split between sets.

This Venn diagram shows 35 students, of whom 19 had been shopping at the weekend, 9 had played football at the weekend and 5 had done both.

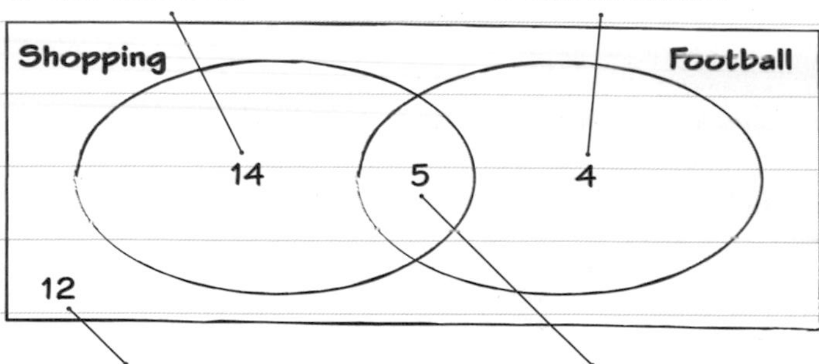

14 + 5 + 4 + 12 = 35 people in total

Each region of a probability Venn diagram represents the probability of a different outcome. The sum of all the probabilities represented in a Venn diagram must equal 1.

Worked example

tier F&H

There are 40 students in a class.
5 study French, German and Spanish.
9 study French and Spanish.
12 study French and German.
7 study German and Spanish.
22 study French.
19 study German.
20 study Spanish.

(a) Draw a Venn diagram. **(3)**

Just French: 22 − (7 + 5 + 4) = 6

Just Spanish: 20 − 11 = 9

Just German: 19 − 14 = 5

None: 40 − (9 + 5 + 2 + 7 + 5 + 4 + 6) = 2

A student is selected at random from the class.

(b) Work out the probability that they study only German. **(1)**

$\frac{5}{40} = \frac{1}{8}$

The diagram shows that 5 students study only German, and there are 40 students in total.

1. Draw the overlapping ovals so you can see which sections you need to complete.
2. Write in the numbers you have been given into the correct sections.
3. Calculate the numbers you haven't been given by subtracting them from the ones you know.
4. Don't forget to calculate the number outside the ovals, for the students who study no languages.

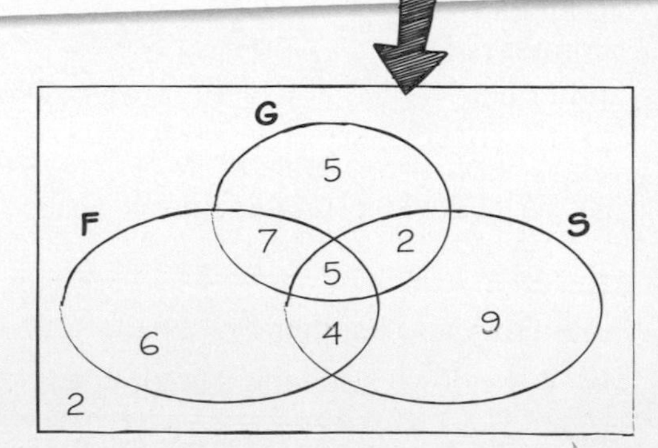

Now try this

The Venn diagram shows the probability a group of students study Science (S), Maths (M) and English (E).

The probability that a student studies English is 0.29.

The probability that a student studies Science is 0.75.

Work out the values of x, y and z. **(4)**

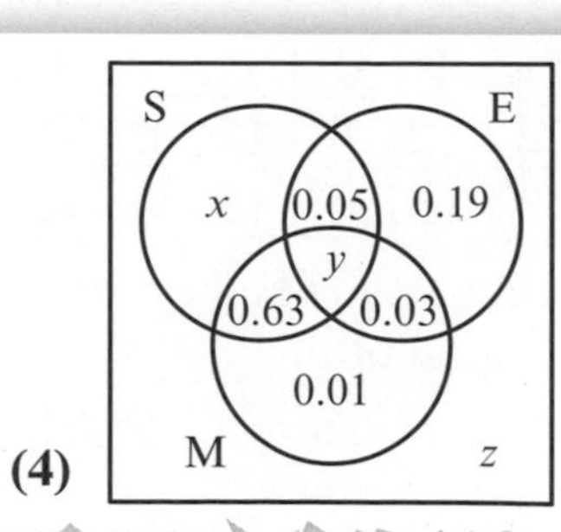

Mutually exclusive and exhaustive events

Mutually exclusive events cannot occur at the **same time**.

Writing probabilities

The probability of rolling a 6 is $\frac{1}{6}$

You can write $P(6) = \frac{1}{6}$

There is one 6. There are six possible outcomes: 1, 2, 3, 4, 5, 6.

The probability of a coin landing heads is $\frac{1}{2}$. You can write $P(\text{head}) = \frac{1}{2}$

There is one head. There are two possible outcomes: heads or tails.

The sum of probabilities

The probabilities of all the possible outcomes of an event add up to 1.

If you know the probability that something will happen, you can calculate the probability that it won't happen.

P(event A doesn't happen)
= 1 − P(event A happens)

This is written as:

P(not A) = 1 − P(A)

Formula for mutually exclusive events

If two events A and B are mutually exclusive then P(A or B) = P(A) + P(B). **LEARN IT!**

This is the **addition law for mutually exclusive events**.
It is sometimes called the 'or' rule.

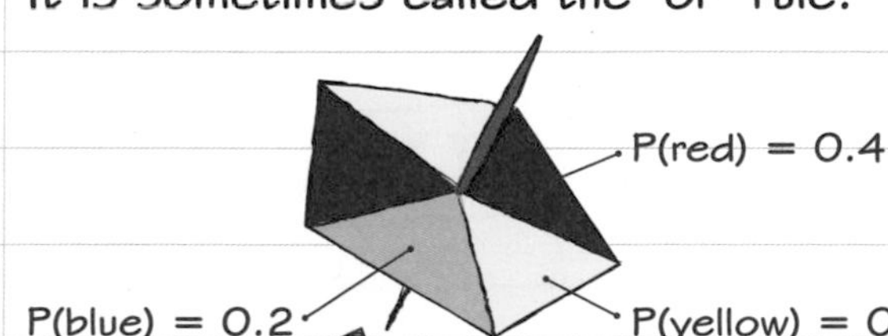

The three outcomes (red, yellow, blue) are **exhaustive** (no others are possible) so their probabilities add up to 1.

This spinner is spun once.
The formula gives further information on the probabilities:

P(red or yellow) = P(red) + P(yellow)
= 0.4 + 0.4 = 0.8

P(red or blue) = P(red) + P(blue)
= 0.4 + 0.2 = 0.6

P(blue or yellow) = P(blue) + P(yellow)
= 0.2 + 0.4 = 0.6

Worked example (tier F&H)

The table gives some information about the probabilities of taking a bead of a given colour from a box.

Colour	Red	Green	Blue	Yellow
Probability	0.1	0.25	x	0.3

(a) Work out the value of x. **(2)**

$x = 1 - (0.1 + 0.25 + 0.3) = 0.35$

(b) Work out the probability that a green bead or a yellow bead is taken. **(1)**

0.25 + 0.3 = 0.55

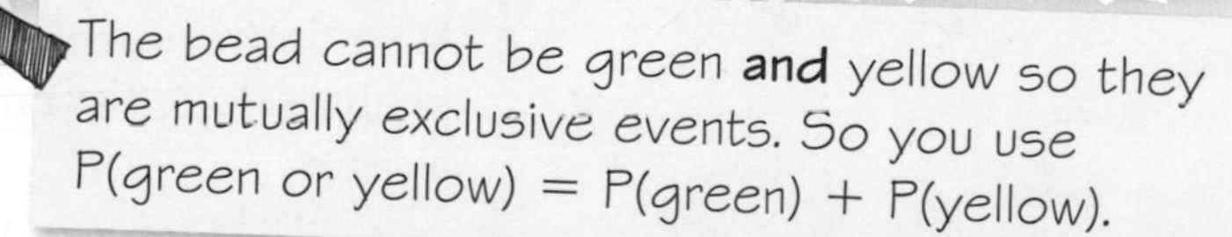

Now try this (tier F&H)

Susie has red, black, blue and pink jumpers in her wardrobe. One day she selects a jumper at random. The table gives information about the probabilities that she selects a jumper of a given colour.

Colour	Red	Black	Blue	Pink
Probability	0.15	0.25	0.32	x

(a) Work out the value of x. **(2)**

(b) Work out the probability that she selects a red jumper or a black jumper. **(1)**

The general addition law

Venn diagrams and probability

LEARN IT!

The shaded area on the Venn diagram represents P(A), which is the probability of A occurring.

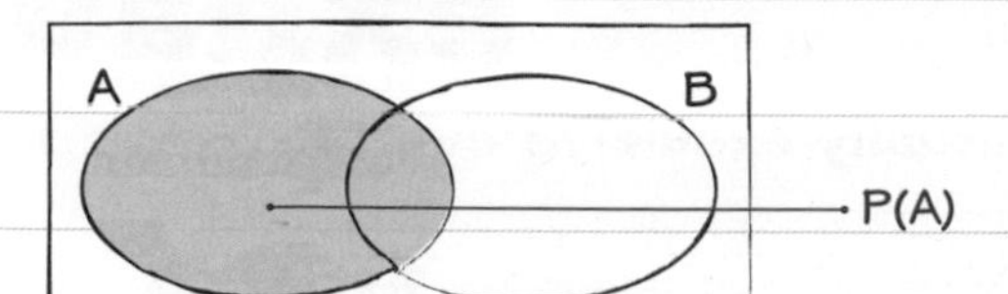

P(A and B) is the probability that **both** A **and** B occur. The shaded area on the diagram represents P(A and B).

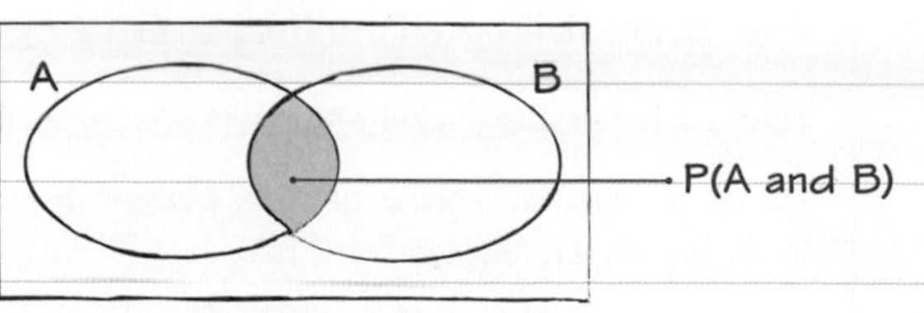

This Venn diagram shows that events A and B are not mutually exclusive, so P(A or B) is the probability that **either** A **or** B occurs or that **both** occur. The shaded area on the diagram represents P(A or B).

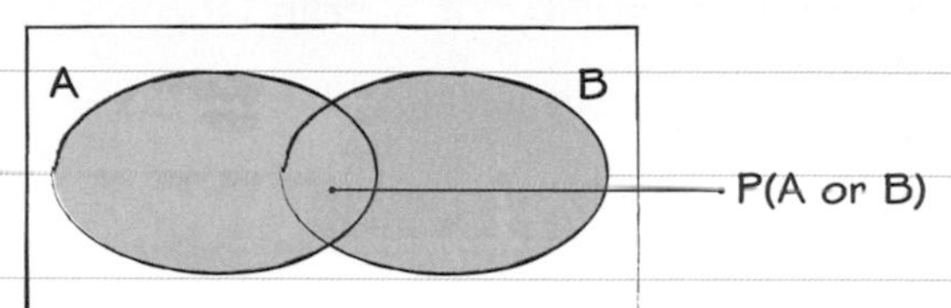

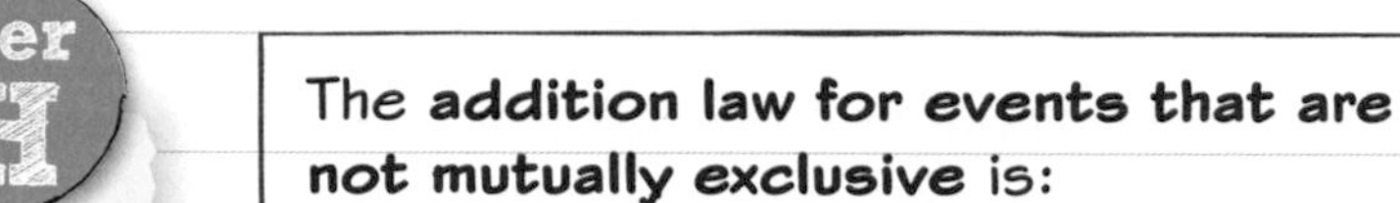

Worked example

tier H

In a group of 50 students, 21 own a smartphone, 15 own a tablet and 6 own both a tablet and a smartphone.
Let S be the event 'own a smartphone'.
Let T be the event 'own a tablet'.
Work out the probability a student owns a smartphone or owns a tablet. **(4)**

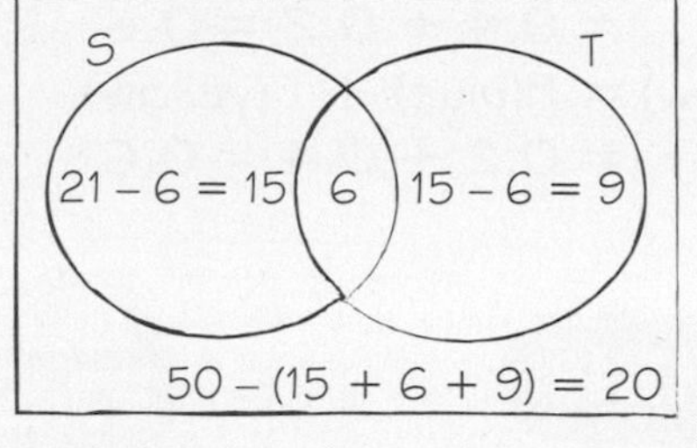

Draw a Venn diagram to help you.

Use the total number of students who own a smartphone and the total number who own a tablet to fill in the empty space on the Venn diagram.

The number who own a smartphone **and** a tablet is the number in the overlap on the Venn diagram = 6

$P(S \text{ or } T) = P(S) + P(T) - P(S \text{ and } T)$

$= \frac{21}{50} + \frac{15}{50} - \frac{6}{50} = \frac{30}{50} \text{ or } \frac{3}{5}$

Use the general addition law.

The **addition law for events that are not mutually exclusive** is:

P(A or B) = P(A) + P(B) − P(A and B)

This is the **general addition law.**

Now try this

In a group of 60 people who are interviewed:

40 have a brother

24 have a sister

8 do not have a brother or a sister.

What is the probability that one of the 60 people selected at random has a brother and a sister? **(4)**

Independent events

If the outcome of one event does not affect the probability of another event occurring, the two events are **independent**.

Formula for independent events

If two events A and B are independent, then P(A and B) = P(A) × P(B). **LEARN IT!**

This is the **multiplication law for independent events**.

A fair dice is rolled twice.

The probability of rolling a 4 and then a 6 is:

$P(4 \text{ then } 6) = P(4) \times P(6) = \frac{1}{6} \times \frac{1}{6} = \frac{1}{36}$

You revised how you could also work out probabilities like these by drawing a **sample space diagram** on page 81.

A fair coin is spun twice.

The probability of getting two heads is:

$P(H \text{ and } H) = P(H) \times P(H) = 0.5 \times 0.5$
$= 0.25$

The outcome of the first spin does not affect the outcome of the second spin, so the events are independent.

Multiple independent events

You can extend the multiplication law for three or more independent events.

If A, B and C are independent events:

P(A and B and C) = P(A) × P(B) × P(C)

The events 'getting a 4' and 'getting a head' are independent. Use P(4 and head) = P(4) × P(head).

tier F&H

Clemmie throws a fair dice and spins a fair coin.

(a) Work out the probability that Clemmie gets a 4 and a head. **(2)**

$\frac{1}{6} \times \frac{1}{2} = \frac{1}{12}$

(b) Work out the probability that Clemmie gets an even number and a head. **(2)**

$\frac{3}{6} \times \frac{1}{2} = \frac{3}{12} = \frac{1}{4}$

Clemmie now throws 2 fair dice and spins a fair coin.

(c) What is the probability that she gets 2 odd numbers and a tail? **(3)**

$\frac{3}{6} \times \frac{3}{6} \times \frac{1}{2} = \frac{9}{72} = \frac{1}{8}$

Now try this

tier F&H

Tom has a spinner with several different coloured sectors of different sizes.
The table gives some information about the probability of getting each colour on the spinner when it is spun once.

Colour	Purple	Blue	Green	Pink
Probability	0.05	x	0.25	0.4

Tom spins the spinner once.

(a) Work out the probability that the spinner stops on blue. **(1)**

Tom spins the spinner twice.

(b) Work out the probability that the spinner stops on pink then green. **(2)**

(c) Work out the probability that the spinner stops on purple both times. **(2)**

Tom spins the spinner three times.

(d) Work out the probability that the spinner stops on purple then green then pink. **(3)**

Tree diagrams

Independent events

The **tree diagram** shows all the possible outcomes when one counter is picked from bag A and one counter is picked from bag B. The outcome of one pick does not affect the outcome of the other pick so the two events are **independent**. There is more about independent events on page 85.

Bag A Bag B

Write the probabilities on the branches.

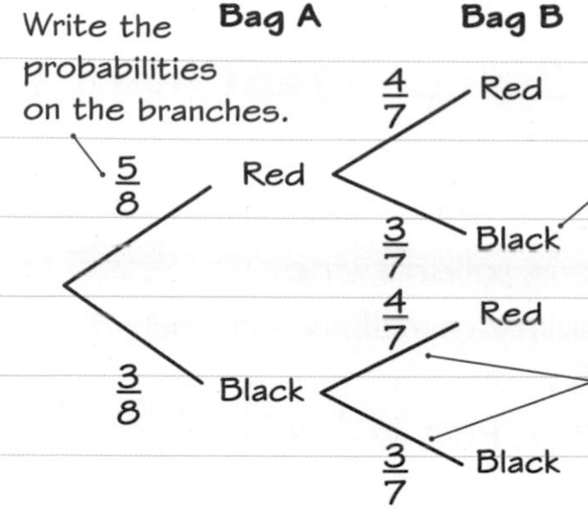

This is the outcome that a red counter is picked from bag A and a black counter is picked from bag B.

On each pair of branches the probabilities add up to 1.

Worked example

Bag X contains 5 black beads and 3 white beads. Bag Y contains 2 black beads and 3 white beads.
Sara takes a bead at random from each bag.
The probability tree diagram gives some information about the probabilities.

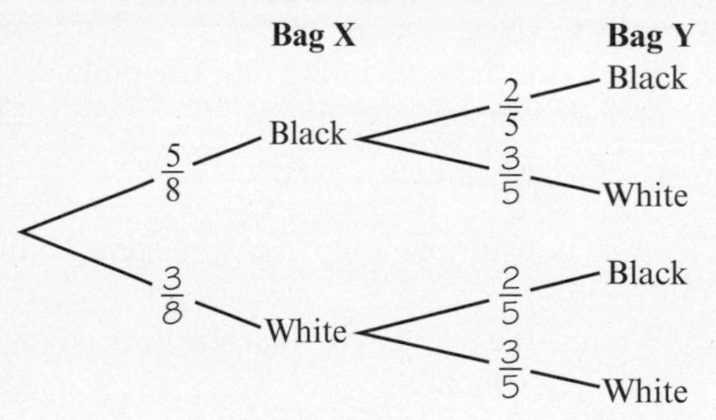

(a) Complete the diagram. **(2)**

(b) Work out the probability that both beads are black. **(2)**

$$\frac{5}{8} \times \frac{2}{5} = \frac{10}{40} = \frac{1}{4}$$

(c) Work out the probability that both beads are the same colour. **(2)**

$$\frac{5}{8} \times \frac{2}{5} + \frac{3}{8} \times \frac{3}{5} = \frac{19}{40}$$

There are two possible outcomes: black, black and white, white. When there is more than one possible outcome from a tree diagram, you **add** the relevant branches to work out the probability that **one** of the events might occur.

Golden rule

To work out probabilities in a tree diagram you:

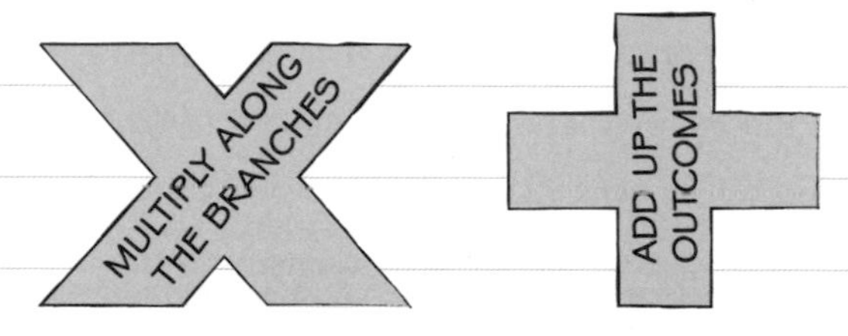

When you **multiply** probabilities, the answer must always be less than 1.

When you **add**, the answer for the sum of probabilities

- for any pair of branches must always be equal to 1
- for all the possible outcomes must always be equal to 1.

You **multiply** along the branches on a tree diagram to work out the probability that **both** events occur.

Now try this

A fair dice is thrown in a game.
A person wins a prize when a 6 is rolled.
Tess rolls the dice three times.

(a) Draw a tree diagram to show this information. **(2)**

(b) Work out the probability that Tess wins three prizes. **(2)**

Conditional probability

You can calculate **conditional probability** from tree diagrams, two-way tables and Venn diagrams.

Calculating conditional probability

A and B are two events.

The probability of event A occurring **given that** event B has already occurred is denoted by P(A|B).

Look out for the key word 'given' – this implies **conditional probability.**

The probability of events A and B **both** occurring is P(A|B) × P(B).

LEARN IT!

Events X and Y are **not** independent, as the outcome of the first pick affects the probabilities for the second pick. This is known as **sampling** (or selection) **without replacement** and always involves conditional probabilities.

Worked example

A bag contains 4 black beads and 3 red beads. One bead is picked at random, not replaced, and then a second bead is picked at random.
X is the event 'the first bead is black'.
Y is the event 'the second bead is black'.

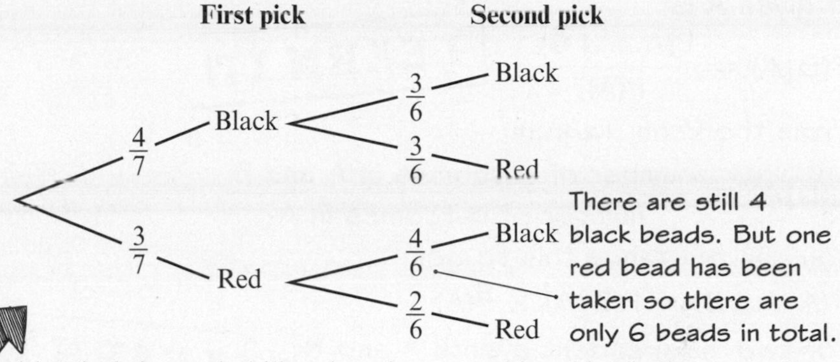

(a) Find the probability that the second bead is black, given that the first bead is black. **(1)**

$P(Y|X) = \frac{3}{6}$

(b) Find the probability that both beads are black. **(2)**

$P(Y|X) \times P(X) = \frac{3}{6} \times \frac{4}{7} = \frac{12}{42} = \frac{2}{7}$

Use the multiplication rule to find probabilities in tree diagrams.

Worked example

tier F&H

The Venn diagram shows probabilities related to two events, A and B.

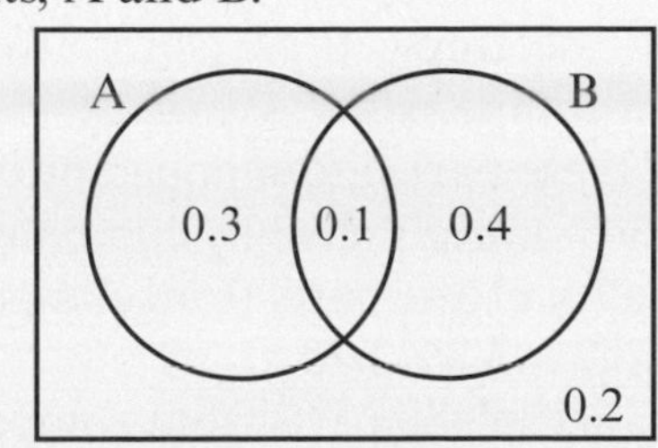

(a) Given that A has happened, calculate the probability that B will happen. **(2)**

$P(B|A) = \frac{0.1}{0.1 + 0.3} = \frac{0.1}{0.4} = \frac{1}{4}$

Find the probability for event A and find the fraction of event A that is the probability for event B.

(b) Given that B has happened, calculate the probability that A will happen. **(2)**

$P(A|B) = \frac{0.1}{0.1 + 0.4} = \frac{0.1}{0.5} = \frac{1}{5}$

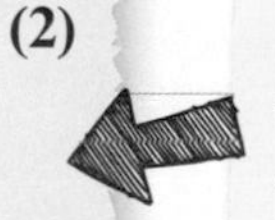

Find the probability for event B and find the fraction of event B that is the probability for event A.

Now try this

There are 6 black beads and 4 red beads in a bag. Two beads are taken at the same time.
Work out the probability that both beads will be the same colour. **(3)**

The formula for conditional probability

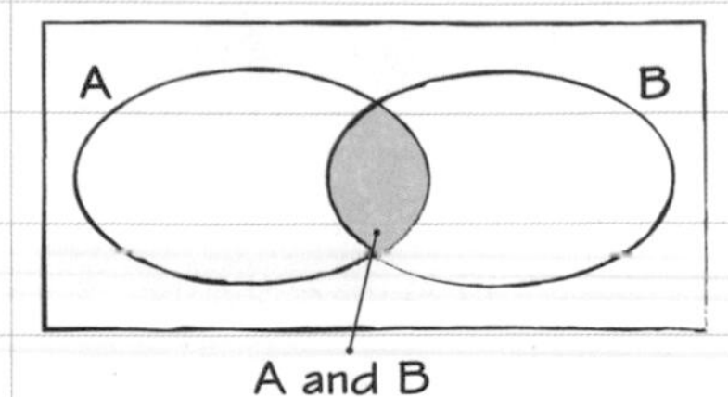

The formula for the conditional probability of B given A is:

$$P(B|A) = \frac{P(A \text{ and } B)}{P(A)}$$

LEARN IT!

From the Venn diagram:

$$P(B|A) = \frac{\text{number of outcomes in A and B}}{\text{number of outcomes in A}}$$

You can rearrange this to get:

P(A and B) = P(B|A) × P(A)

For two independent events A and B:

P(A|B) = P(A) and P(B|A) = P(B)

LEARN IT!

Worked example

tier F&H

A and B are two events such that P(A) = 0.5 and P(A and B) = 0.4

Given that A and B are independent events, work out P(B). **(2)**

$$P(B|A) = \frac{P(A \text{ and } B)}{P(A)} = \frac{0.4}{0.5} = \frac{4}{5} = 0.8$$

A and B are independent so

P(B|A) = P(B) = 0.8

If A and B are independent, the fact that A has happened has no effect on the probability of B happening.

Worked example

tier F&H

Kim has a set of 20 tiles. 10 of the tiles have the letter X on them, of which 6 are red. The other tiles have the letter Y on them, of which 7 are red. All other tiles are white. Kim takes a single tile at random.

Let A be the event 'a white tile is selected'. Let B be the event 'a tile with the letter X is selected'.

(a) Find P(A). **(3)**

(10 − 6) + (10 − 7) = 4 + 3 = 7 white tiles out of a total of 20 tiles

$$P(A) = \frac{7}{20}$$

(b) Find P(B|A). **(2)**

$$P(B|A) = \frac{\text{number of outcomes in A and B}}{\text{number of outcomes in A}} = \frac{4}{7}$$

(c) Find P(A and B). **(2)**

$$P(A \text{ and } B) = P(B|A) \times P(A) = \frac{4}{7} \times \frac{7}{20} = \frac{4}{20} = \frac{1}{5}$$

There are 6 red tiles out of 10 Xs and 7 red tiles out of 10 Ys. So that leaves 4 white X tiles and 3 white Y tiles.

Use the formula for conditional probability. Make sure you write it out before substituting the values.

Now try this

tier F&H

1 JLD Engineering gets parts from two different companies.
70% of these parts are supplied by company A and 30% are supplied by company C.
The quality of the parts from company A is 97% good and 3% bad.
The quality of the parts from company C is 95% good and 5% bad.
Let A denote the event 'a part from company A'. Let G denote the event 'good quality'.
(a) Draw a probability tree diagram. **(3)**
(b) Find (i) P(A), (ii) P(G|A), (iii) P(A and G), (iv) P(G). **(4)**

2 Events A and B are such that P(A) = 0.4, P(B) = 0.5 and P(A and B) = 0.2. Show that A and B are independent events. **(2)**

Index numbers

Simple index numbers are a way of tracking changes in value through time.

Calculating an index number

Index numbers are like percentages which describe changes in costs or prices from year to year.

The cost or prices are compared with a **base year price**.

In the **base year** the index number is defined as 100.

You can calculate a simple index number for year n using the rule:

$$\text{Index} = \frac{\text{cost in year } n}{\text{cost in base year}} \times 100$$

LEARN IT!

2010

The index number of the laptop cost in 2010 (the base year) is defined as 100.

2016

In 2016, the index number is $\frac{425}{500} \times 100 = 85$

An index number more than 100 means that the price in year n is higher than in the base year.

If a price goes down then the index number will be less than 100 in year n.

Using an index number

You can use index numbers to work out costs and prices.

Use the formula:

$$\text{Cost in year } n = \text{cost in base year} \times \frac{\text{new index number}}{100}$$

The value of an investment in 2005 (the base year) was £4000 and the index number for 2017 was 108.

So the value of the investment in 2017 was £4000 × $\frac{108}{100}$ = £4320

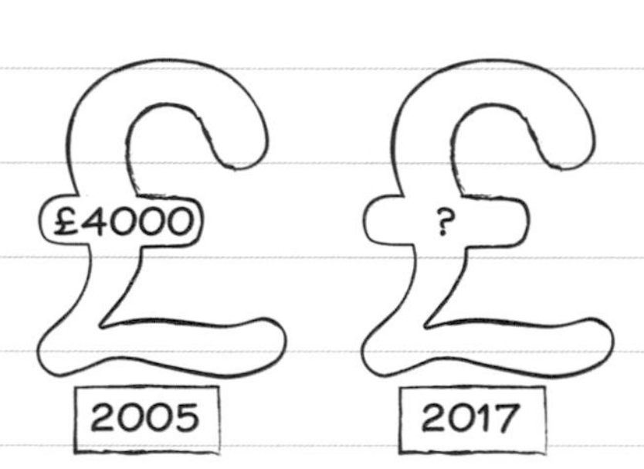

Worked example

tier F&H

The cost of some groceries in June 2017 was £120.

The cost of the same groceries in July 2017 was £121.80.

June 2017 is the base month.

(a) Work out the index number for the cost of the groceries in July 2017. **(2)**

July index number = $\frac{121.80}{120} \times 100 = 101.5$

The index number for the cost of the same groceries in August 2017 is 98.

(b) Work out the cost of the groceries in August 2017. **(2)**

Cost in August = $120 \times \frac{98}{100}$ = £117.60

In calculations involving index numbers you will be told which year (or month) to use as the base.

Remember that when you have to **find** an index number you use:

$$\frac{\text{cost in year } n}{\text{cost in base month}} \times 100$$

When you have to **use** an index number you use:

$$\text{cost in base month} \times \frac{\text{new index number}}{100}$$

Now try this

In 1999, the national minimum wage for an adult was £3.60 per hour. In 2015, it was £6.70 per hour.

(a) Using 1999 as the base year, work out the index number for the adult minimum wage in 2015. **(2)**

In 1999, the national minimum wage for young workers was £3.00 per hour. In 2015, the index number for the national minimum wage for young workers was $176\frac{2}{3}$.

(b) Work out the national minimum wage for young workers in 2015. **(2)**

Had a look ☐ Nearly there ☐ Nailed it! ☐

RPI and CPI

You need to be able to use different types of index numbers in context. Two common index numbers are the **Retail Price Index (RPI)** and the **Consumer Prices Index (CPI)**.

Retail Price Index (RPI)

The Retail Price Index shows changes in the cost of living. It gives a measure of prices in everyday life, such as rental payments, food, heating and travel costs. The Government uses the RPI to set interest rates for student loans.

The RPI for all items is calculated from base year 1987 with an index of 100.

The table shows the annual average RPI from 2006 to 2016.

Year	2006	2007	2008	2009	2010	2011	2012	2013	2014	2015	2016
RPI	198.1	206.6	214.8	213.7	223.6	235.2	242.7	250.1	256.0	258.5	263.1

Source: Office for National Statistics

The percentage increase in the RPI from 2006 to 2016 is $\frac{263.1 - 198.1}{198.1} \times 100 = 32.8\%$

Consumer Prices Index (CPI)

The Consumer Prices Index also shows changes in the cost of living, but does not include mortgage payments. The CPI is used to index benefits, tax credits and pensions in the UK.

The CPI for all items is calculated from base year 2015 with an index of 100.

The table shows the annual average CPI from 2006 to 2016.

Year	2006	2007	2008	2009	2010	2011	2012	2013	2014	2015	2016
CPI	79.9	81.8	84.7	86.6	89.4	93.4	96.1	98.5	100.0	100.0	100.7

Source: Office for National Statistics

There was no increase in the CPI from 2014 to 2015.

Worked example

tier F&H

The table shows information about the CPI and the price of a rail season ticket from Manchester to Macclesfield from 2015 to 2017.

	2015	2016	2017
CPI	100	100.7	103.4
Price of season ticket (£)	1968	1988	2024

Sources: Office for National Statistics and National Rail

(a) Describe how the increase in price of a rail season ticket from Manchester to Macclesfield compares with the CPI from 2015 to 2017. **(3)**

Price in 2016 as a percentage of price in 2015 is $\frac{1988}{1968} \times 100 = 101.01$ (2 d.p.)

This is above the CPI.

Price in 2017 as a percentage of price in 2016 is $\frac{2024}{1988} \times 100 = 101.81$ (2 d.p.)

This is below the CPI.

The CPI in 2012 was 96.1.

(b) Work out what the price of a rail season ticket would have been in 2012. **(2)**

$$\text{Index in 2012} = \frac{\text{cost in 2012}}{\text{cost in 2015}} \times 100$$

$$\text{so } 96.1 = \frac{\text{cost in 2012}}{1968} \times 100$$

$$\text{Cost in 2012} = \frac{96.1 \times 1968}{100} = £1891.25$$

Now try this

tier F&H

In the base year the RPI is 100.

The base year for the Retail Price Index (RPI) in the UK is 1987.
In August 2017 the RPI was 274.7.
The average house price in the UK in 1987 was £40 000. In 2017 the average house price was £223 257.
Compare the increase in the average house price from 1987 to 2017 with the RPI. **(2)**

GDP

Another common index number is the **Gross Domestic Product (GDP)**.

Gross Domestic Product is the main measure of economic output based on the value of goods and services produced by a country or region in a given time period.

GDP is an indicator for inflation and is often used to compare the economies of different countries.

The table shows the GDP quarterly growth for the UK as a percentage.

This means Quarter 1 of 2015, i.e. January to March 2015.

Year	2015				2016				2017			
Quarter	1	2	3	4	1	2	3	4	1	2	3	4
%	0.3	0.6	0.4	0.7	0.2	0.5	0.5	0.7	0.3	0.2	0.5	0.4

Source: Office for National Statistics

Worked example

tier F&H

The table shows the percentage changes in UK GDP for Quarter 4 of 2017.

Sector of the UK economy	Percentage change from previous quarter
Total GDP	0.5
Agriculture	−0.4
Production	0.6
Manufacturing	1.3
Construction	−1.0
Services	0.6

Source: Office for National Statistics

(a) Describe the percentage change in the GDP for the Agriculture sector in the last quarter of 2017. **(1)**

It decreased by 0.4%.

In the last quarter of 2016, the percentage change for the Production sector was the same as the percentage change for the Services sector. A government minister says 'This means both sectors must have increased their contribution to GDP by the same amount'.

(b) Is this statement true or false? Explain your answer. **(2)**

Not possible to say, because you do not know the contribution of each sector.

If the contribution of one sector is larger than the other, then 0.6% of the larger amount is greater than 0.6% of the smaller amount.

Now try this

tier F&H

The table shows the quarter-on-previous-quarter GDP growth for the UK, USA and Japan in 2017.

	Q1	Q2	Q3	Q4
UK	0.3	0.2	0.5	0.4
USA	0.3	0.8	0.8	0.6
Japan	0.3	0.6	0.6	0.1

Source: Office for National Statistics

(a) Identify the greatest percentage increase and decrease in GDP growth for these economies during 2017. **(2)**

(b) Which was the most consistent economy during 2017? Explain your answer. **(2)**

Weighted index numbers

You may need to measure how a set of items changes in value. You can do this using a **weighted index number**. Each item is assigned a different weight in the calculation to show how important it is.

The Consumer Prices Index (CPI) and the Retail Price Index (RPI) (see page 90) are the most well-known weighted index numbers.

Calculating weighted index numbers

You can calculate the weighted index number for a set of items by using the formula for the weighted mean:

$$\text{Weighted mean} = \frac{\sum(\text{value} \times \text{weight})}{\sum \text{weights}} \text{ or } \bar{x} = \frac{\sum xw}{\sum w}$$

where w represents the weight of each item.

Worked example

tier H

Shona invests in shares. In 2015 she owned 1000 shares in company A and 2500 shares in company B. She still had these investments in 2017.

The weight for company A is 1000 and for company B is 2500.

	2015	2017
Company A	£2.50	£3.00
Company B	£2.80	£4.00

(a) Work out the weighted index number for Shona's shares in 2017 taking 2015 as the base year. **(3)**

$$\text{Weighted mean (2015)} = \frac{2.50 \times 1000 + 2.80 \times 2500}{1000 + 2500} = £2.71 \text{ (3 s.f.)}$$

$$\text{Weighted mean (2017)} = \frac{3 \times 1000 + 4 \times 2500}{1000 + 2500} = £3.71 \text{ (3 s.f.)}$$

The weighted mean for each year will be between the smallest and largest values. This is a useful check for your answers.

Index number for 2017 taking 2015 as base year

$$= \frac{3.71}{2.71} \times 100 = 136.90... = 136.9 \text{ (1 d.p.)}$$

(b) Interpret your answer to part (a). **(1)**

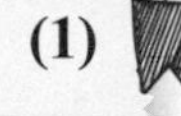

Percentage increase from 2015 to 2017 = 136.9 − 100 = 36.9%

Use the rule:
weighted index number

$$= \frac{\text{current weighted mean price}}{\text{base year weighted mean price}} \times 100$$

Now try this

tier H

The table gives information about the profits per item from two factories in 2016 and 2017.

Three times as many items come from Factory A as from Factory B.

Using 2016 as the base year, work out the weighted index number for the profits for 2017. **(3)**

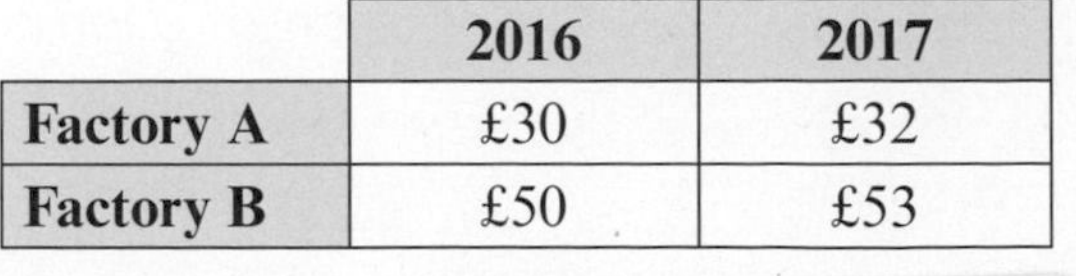

	2016	2017
Factory A	£30	£32
Factory B	£50	£53

Use a weight of 3 for Factory A.

Chain base index numbers

A **chain base** index number compares **this** year's value with **last** year's value (the **base**).
The calculation is very similar to finding a **simple** index number except that the base changes every year.

$$\text{Chain base index number} = \frac{\text{value this year}}{\text{value last year}} \times 100$$

LEARN IT!

Using a chain base index number

You can use chain base index numbers to find a new price.

Price this year =

$$\frac{\text{chain base index number for this year}}{100} \times \text{price last year}$$

If this year's chain base index number is 120, then this year's price is $\frac{120}{100}$ × last year's price.

This means the price is 20% higher than last year.

CPI can be given as a chain base index number. If the chain base index number of annual CPI in 2012 was 102.8, this means that prices rose by 2.8% on average between 2011 and 2012.

Chain base vs simple index numbers

Chain base and simple index numbers are always worked out from

$$\frac{\text{value}}{\text{value in the base year}} \times 100$$

- For **simple**, the base year does not change.
- For **chain base**, the base year is always the previous year.

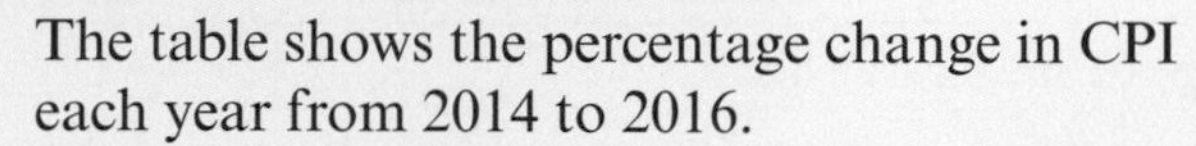

The table shows the percentage change in CPI each year from 2014 to 2016.

Year	2014	2015	2016
Percentage change in CPI	1.5	0	0.7

(a) Write down the chain base index number from 2015 to 2016. **(1)**

At the close of 2014 the price of gold was \$1199.25 per ounce.

At the close of 2016 the price of gold was \$1174 per ounce.

(b) Explain whether the price of gold increased in line with the CPI between 2014 and 2016. **(2)**

(a) The CPI rose by 0.7% in 2015.
The chain base index number of 100 + 0.7 = 100.7

(b) Using the chain base index number:
In 2015, the price of gold should be the same as in 2014, because the % change in CPI is 0.
In 2016, the price of gold should be

$$\frac{100.7}{100} \times \$1199.25$$
$$= \$1207.64 \text{ per ounce}$$

However, the actual price of gold decreased from 2014 to 2016 so its price does not change in line with CPI.

Now try this

tier H

The table shows the price of one model of car for three years.

Year	2016	2017	2018
Price (£)	18 500	18 900	19 300

(a) Using 2016 as the base year, find the chain base index numbers for 2017 and 2018. **(3)**

(b) Explain why the two index numbers are not the same, even though the price rise is £400 in each year. **(1)**

Had a look ☐ Nearly there ☐ Nailed it! ☐

Crude rates

Crude rates are a simple way to compare population statistics such as births, deaths and employment levels.

- The **crude birth rate** is the number of births per thousand of the population.
- The **crude death rate** is the number of deaths per thousand of the population.

Crude rates can be calculated using this general formula:

$$\text{Crude rate} = \frac{\text{number of (deaths/births/people unemployed)} \times 1000}{\text{total population}}$$

In the exam, the formula you need to use will be given in the question.

In a town with a population of 63 400, 5700 people are unemployed.
You can work out the crude unemployment rate per thousand of the population.

$$\text{Crude unemployment rate} = \frac{\text{number of people unemployed} \times 1000}{\text{total population}}$$

$$= \frac{5700 \times 1000}{63400}$$

$$= 89.9 \text{ (1 d.p.) unemployed per 1000}$$

You might be asked to work out crude rate of change per hundred in a question instead of per thousand. Just replace 1000 by 100 in the formula.

Worked example

tier F&H

The table shows the numbers of live births in England and Wales in 2016 by age group of mother, and the total population in each age group.

	Age of mother at birth			
	20 to 24	**25 to 29**	**30 to 34**	**35 to 39**
Number of live births	102 607	196 132	220 129	125 205
Number in age group (thousands)	1837	1986	1966	1871

Source: Office for National Statistics

(a) What was the crude birth rate per 1000 people for the age group 30 to 34 in England and Wales in 2016? **(2)**

$$\text{Crude birth rate for age 30 to 34} = \frac{\text{number of live births to mothers aged 30 to 34} \times 1000}{\text{total female population aged 30 to 34}}$$

$$= \frac{220129 \times 1000}{1966 \times 1000} = 112 \text{ (3 s.f.) per 1000}$$

The total number of live births in England and Wales in 2016 was 696 271. The total female population was 29 546 000.

(b) What was the crude birth rate for the female population in England and Wales in 2016? **(2)**

$$\text{Crude birth rate for female population} = \frac{\text{number of live births} \times 1000}{\text{total female population}}$$

$$= \frac{696271 \times 1000}{29546000} = 23.6 \text{ (3 s.f.) per 1000}$$

Now try this

tier F&H

From October to December 2017 there were 32.15 million people employed in the UK and 1.47 million people unemployed. Work out the crude unemployment rate for the population of working age. **(3)**

Standardised rates

Crude rates do not take into account the age of a population. In a town with a large elderly population the crude death rate is likely to be high, so you can make use of a **standard population** calculation to enable valid comparisons between distributions.

$$\text{Standard population} = \frac{\text{number in age group} \times 1000}{\text{total population}}$$

$$\text{Standardised rate} = \frac{\text{crude rate}}{1000} \times \text{standard population}$$

Worked example

tier H

The table shows a breakdown of the deaths in one year for a town with a total population of 45 980.

Age group	Number in group	Number of deaths	Standard population	Crude death rate	Standardised death rate
0–19	3215	42	$\frac{3215 \times 1000}{45980}$ $= 69.92$	$\frac{42 \times 1000}{45980}$ $= 0.91$	$\frac{0.91}{1000} \times 69.92$ $= 0.06$
20–35	15 462	840	$\frac{15462 \times 1000}{45980}$ $= 336.3$	$\frac{840 \times 1000}{45980}$ $= 18.23$	$\frac{18.23}{1000} \times 336.3$ $= 6.13$
36–65	20 436	2103	$\frac{20436 \times 1000}{45980}$ $= 444.5$	$\frac{2103 \times 1000}{45980}$ $= 45.73$	$\frac{45.73}{1000} \times 444.5$ $= 20.33$
>65	6867	2497	$\frac{6867 \times 1000}{45980}$ $= 149.3$	$\frac{2497 \times 1000}{45980}$ $= 54.31$	$\frac{54.31}{1000} \times 149.3$ $= 8.11$

(a) Find and compare the standardised death rates for each age group. **(4)**

The 0–19 age group has the lowest standardised death rate and the 35–65 age group has the highest standardised death rate.

(b) Give an interpretation of what these values mean. **(1)**

The standardised death rates give the number of deaths per 1000 of the population for each age group.

Now try this

tier H

Abdul is comparing death rates for the population aged over 65 in two towns, A and B.
He has this information.

Town	Standard population (age >65)	Crude death rate (age >65)
A	152	60.5
B	141	72.3

He thinks the standardised death rate for the population aged over 65 will show there are just over 2 more deaths per 1000 in town A than in town B.

Is Abdul correct? Explain your answer. **(3)**

Binomial distributions 1

The **probability distribution** of X is the set of values X can take and the associated probabilities.

Probability distributions are a way of showing all the outcomes and their probabilities. The outcomes have to be **numerical,** such as the number of heads when three fair coins are spun.

X is the number of heads

X	0	1	2	3
P(X)	0.125	0.375	0.375	0.125

The probabilities **must** add up to 1

The probability of getting exactly 1 head is 0.375

Binomial distribution

A **binomial distribution** occurs when there are only two possible outcomes for an experiment, where one outcome can be classed as **success** and one as **failure.**

If you throw a dice and count getting a 6 as a success, then getting 1, 2, 3, 4 or 5 is failure.

Let P(6) = p and P(not 6) = q.

The probability tree diagram shows the outcomes for throwing two dice. Consider the two throws of the dice as trials.

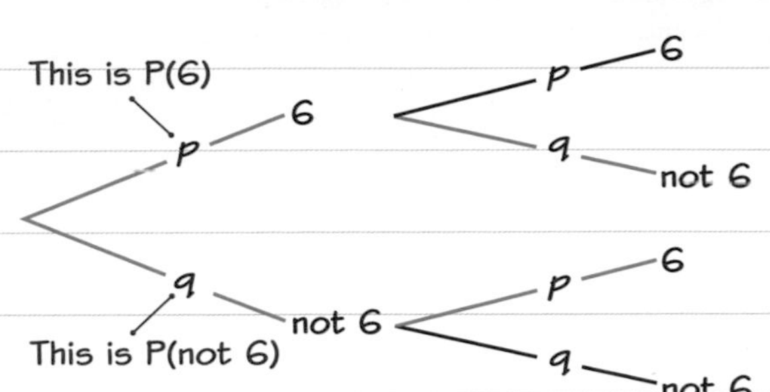

The blue lines in the probability tree show the two ways of getting one '6', and one 'not 6' with total probability $p \times q + q \times p = 2pq$.

Since success or failure are the only possible outcomes in a single trial, $p + q = 1$

Probability and binomials

You can use the expansion of

$(p + q)^2 = p^2 + 2pq + q^2$

to work out probabilities when there are two repeated trials of two events that are mutually exclusive.

Worked example

tier H

Two bags, A and B, each contain 5 white and 3 red beads.
A bead is taken at random from bag A and a bead is taken at random from bag B.

The conditions for the two bags are **identical.**

(a) Find the probability that both beads are white. **(2)**

Let getting white be a success.

$p = \frac{5}{8}$ and $q = \frac{3}{8}$

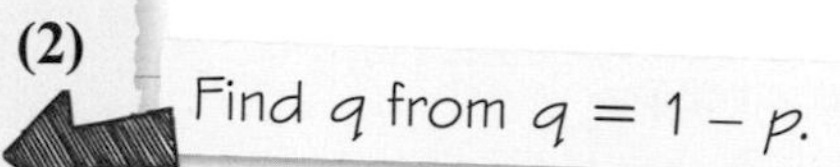

Probability of two white beads $= p^2 = \frac{25}{64}$

The probability of two successes out of two is p^2.

(b) Find the probability that exactly one bead is white. **(2)**

Probability $= 2 \times \frac{5}{8} \times \frac{3}{8} = \frac{30}{64}$

The probability of one success and one failure out of two trials is $2pq$.

Now try this

Alice has a biased dice. The probability of getting 6 on any roll is 0.2.
Alice rolls the dice twice.

(a) Work out the probability that she gets exactly one 6. **(2)**

(b) Work out the probability that she does not get a 6 in either of the rolls. **(2)**

Binomial distributions 2

When the number of trials is more than two it becomes more difficult to draw probability tree diagrams.

Instead, you can use the **binomial expansion** to find the probability distribution.

For three, four and five trials, the expansions are:

$(p + q)^3 = p^3 + 3p^2q + 3pq^2 + q^3$ ← 3 trials

$(p + q)^4 = p^4 + 4p^3q + 6p^2q^2 + 4pq^3 + q^4$ ← 4 trials

$(p + q)^5 = p^5 + 5p^4q + 10p^3q^2 + 10p^2q^3 + 5pq^4 + q^5$ ← 5 trials

You do not have to remember any of these formulae.
In the exam, the one you need to use will be given in the question.

You can find the coefficients of a binomial expansion using the ${}_nC_r$ function on your calculator. For example, to find the coefficients of each of the 6 terms in the expansion of $(p + q)^5$ use $n = 5$ and $r = 0, 1, 2, 3, 4, 5$

The notation $B(n, p)$ denotes a binomial distribution with n trials and probability of success p.

For the binomial distribution $B(n, p)$:

- the probability of failure is q, where $q = 1 - p$
- the probabilities for the events of n binomial trials are the terms of the expansion of $(p + q)^n$
- the mean is np.

Golden rule

The binomial distribution is a suitable model if:

- the number of trials is fixed
- the trials are independent
- there are two possible outcomes for each trial (success and failure).

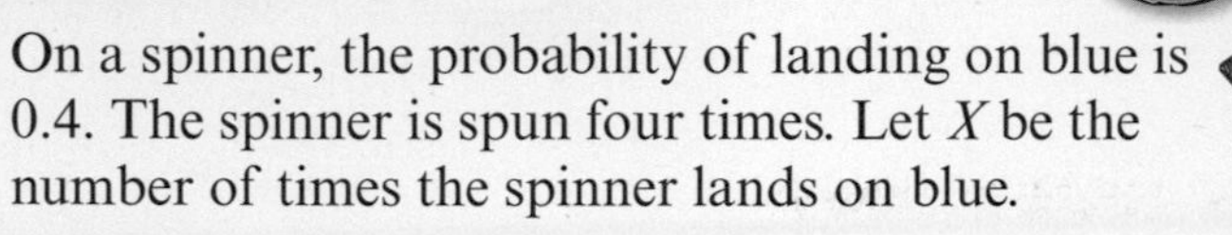

Worked example

tier H

On a spinner, the probability of landing on blue is 0.4. The spinner is spun four times. Let X be the number of times the spinner lands on blue.

(a) Work out $P(X = 2)$. **(2)**

$p = 0.4$, so $q = 0.6$

$P(X = 2) = 6 \times 0.4^2 \times 0.6^2 = 0.3456$

(b) Work out $P(X > 2)$. **(2)**

$P(X > 2) = 4 \times 0.4^3 \times 0.6 + 0.4^4$

$= 0.1536 + 0.0256 = 0.1792$

(c) Estimate the mean number of times the spinner will land on blue in 100 spins. **(1)**

For 100 spins, mean number of times for success $= 100 \times 0.4 = 40$

This is B(4, 0.4). Use: $(p + q)^4 = p^4 + 4p^3q + 6p^2q^2 + 4pq^3 + q^4$

$P(X = 2)$ means the number of successes is 2. Use the term which has the same power of p as the number of successes, i.e. $6p^2q^2$.

$X > 2$ means that $X = 3$ or $X = 4$. These two cases are mutually exclusive so the probabilities can be added. Use the terms in the expansion which include p^3 and p^4 i.e. p^4 and $4p^3q$.

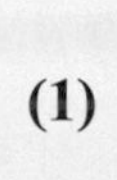

Now try this

tier H

On the road by Ken's house the probability that any car passing is an SUV is 0.15.
Let X be the number of SUVs in the first 5 cars that pass Ken's house.

(a) Work out the probability that all 5 cars are SUVs. **(1)**

(b) Work out the probability that at least 4 cars are SUVs. **(2)**

(c) Estimate the mean number of SUVs for the first 500 cars that pass Ken's house. **(2)**

 Had a look ☐ Nearly there ☐ Nailed it! ☐

Normal distributions

The **normal distribution** is a suitable model if:

- the data is continuous
- the distribution is symmetrical and bell-shaped
- the mode, median and mean are approximately equal.

Distribution of values

In a normal distribution:

- 68% of values are within $\pm\sigma$ of the mean μ
- 95% of values are within $\pm 2\sigma$ of the mean μ
- 99.7% of values are within $\pm 3\sigma$ of the mean μ

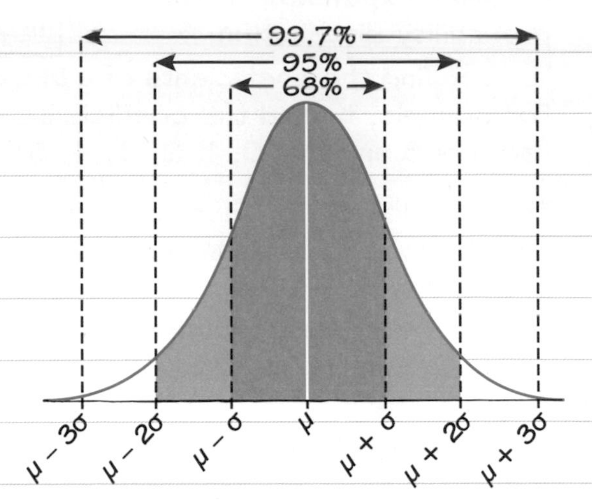

The number of standard deviations, σ, of a value from the mean can be worked out using:

$$\text{number of sds from mean} = \frac{\text{value} - \text{mean}}{\text{standard deviation}}$$

Worked example

tier H

The heights of a species of daffodil are normally distributed. 2.5% of the heights are greater than 16.5 cm. 50% of the heights are greater than 13.5 cm.

(a) Find the mean and the standard deviation. **(2)**

$\mu = 13.5$

$\mu + 2\sigma = 16.5$ so $\sigma = \frac{16.5 - 13.5}{2} = 1.5$

(b) Work out the probability that the heights of the daffodils are greater than 18 cm. **(2)**

Number of standard deviations from the mean $= \frac{18 - 13.5}{1.5} = 3$

Probability $= \frac{100 - 99.7}{2} = 0.15\%$

> The distribution is symmetrical about the mean so 50% must be above the mean.

> 95% lie within $\mu + 2\sigma$ and $\mu - 2\sigma$ so 2.5% are more than $\mu + 2\sigma$.

> 99.7% of the heights lie between $\mu + 3\sigma$ and $\mu - 3\sigma$ so the probability that the height is greater than 18 cm is: $\frac{(100 - 99.7)}{2} = 0.15\%$

Now try this

tier H

The masses of adult trout are normally distributed with mean 2.4 kg and standard deviation 0.2 kg.

(a) Work out the mass which is exceeded by 2.5% of this trout population. **(2)**

(b) Between what values would almost all of the trout masses lie? **(2)**

(c) Draw a sketch of the distribution. **(2)**

(d) What is the probability that the weight of an adult trout will be less than 2.2 kg? **(2)**

> When sketching a normal distribution, sketch to 3 standard deviations either side of the mean.

Standardised scores

Standardised scores are used to compare values from different frequency distributions.

Understanding standardised scores

When students do different tests it is not always fair to compare the marks, as one test might have been more difficult than another.

You can use the rule: **LEARN IT!**

$$\text{Standardised score} = \frac{\text{mark} - \text{mean}}{\text{standard deviation}} = \frac{x - \mu}{\sigma}$$

to compare marks on different tests.

A mark **greater** than the mean gives a **positive** standardised score.

A mark **lower** than the mean gives a **negative** standardised score.

The **higher** the standardised score, the **better** the student's performance on the test.

Bigger standard deviations result in lower standardised scores.

The diagrams show the distributions of marks in two tests.

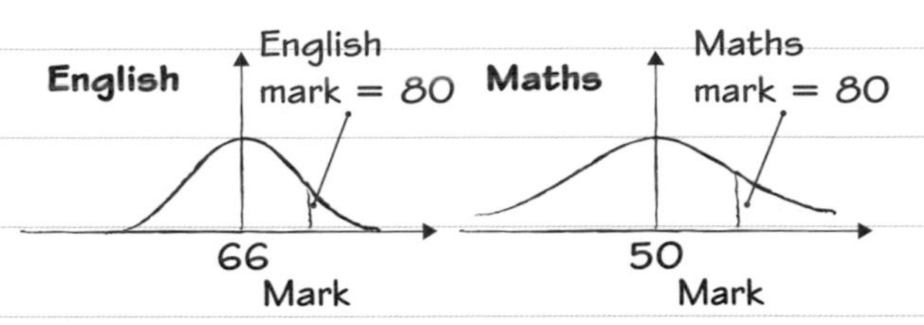

English: $\mu = 66$; $\sigma = 16$.

Maths: $\mu = 50$; $\sigma = 30$.

A mark of 80 in English gives a standardised score of $\frac{80 - 66}{16} = 0.875$

A mark of 80 in Maths gives a standardised score of $\frac{80 - 50}{30} = 1.0$

The standardised score for a mark of 80 in Maths is higher than for 80 in English, showing that the performance in Maths was better.

Worked example

tier H

50 people took an exam. The mark, x, for each person was recorded.

$\sum x = 3100$, $\sum x^2 = 203\,450$

Maisie got a mark of 86. Work out her standardised score. **(4)**

$$\text{Mean} = \frac{3100}{50} = 62$$

$$\text{Standard deviation} = \sqrt{\frac{203\,450}{50} - \left(\frac{3100}{50}\right)^2}$$

$$= \sqrt{225} = 15$$

$$\text{Standardised score} = \frac{86 - 62}{15} = 1.6$$

You are given $\sum x$ and $\sum x^2$, so use this formula:

$$\text{standard deviation} = \sqrt{\frac{\sum x^2}{n} - \frac{(\sum x)^2}{n}}$$

Use the formula for standardised score:

$$\frac{\text{Maisie's mark} - \text{mean}}{\text{standard deviation}}$$

Now try this

tier H

1 Abdul took two tests in an interview for a job. On the aptitude test he got 48 marks. On the skills test he got 64 marks. The table gives information about the two tests.

	Mean mark	Standard deviation
Aptitude test	45	10
Skills test	55	8

Work out Abdul's standardised scores on the two tests. **(3)**

2 100 people took a test. The variable x denotes the mark they each got.

$\sum x = 4600$, $\sum x^2 = 226\,000$

(a) Shola got 40 marks. Work out his standardised score. **(3)**

(b) Alice had a standardised score of 1.5. Work out how many marks she got. **(2)**

 Had a look ☐ Nearly there ☐ Nailed it! ☐

Quality assurance and control charts 1

When items are made on a production line, the company needs to check their **quality**. One way of doing this is to take a sample periodically and test the items in the sample – by weighing or by measuring lengths, for example.

Control charts and sample means

A tester will work out the **sample mean** of the measurements and plot it on a **quality control chart**.

Each sample will have a different mean but the distribution of the sample means is generally normal with mean μ.

- **Warning limits** are $\pm 2\sigma$ from the target mean μ.
- **Action limits** are $\pm 3\sigma$ from the target mean μ.

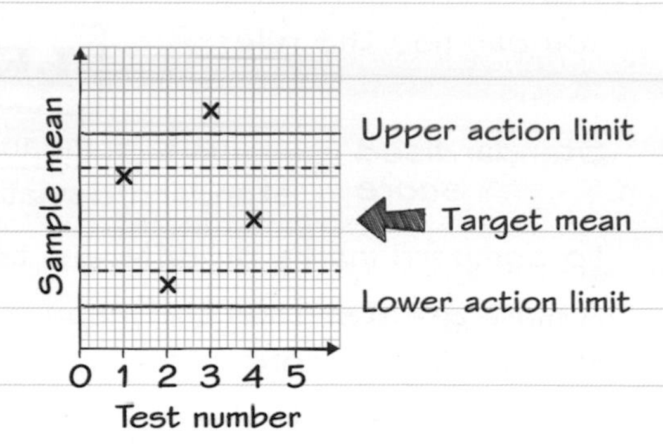

Actions to ensure quality control

1. The sample mean is within the warning limits – do nothing.
2. The sample mean is between the warning limit and the action limit – do a second test immediately.
3. The sample mean is outside the action limit. Stop the machine, check it and reset.

LEARN IT!

If everything is working correctly:

- values of the sample mean should lie within $\pm 2\sigma$ of the target mean μ, in 19 out of 20 samples (95%)
- values of the sample mean should lie within $\pm 3\sigma$ of the target mean in 998 out of 1000 samples (99.7%), so in this case the probability of being outside these limits is 0.2% (assuming the machine is working properly).

Worked example

tier H

In order to check the lengths L mm of precision needles, a factory tests samples of 10 needles every 30 minutes. For the first three samples, the values were

1. $\sum L = 180$ **2.** $\sum L = 183.5$ **3.** $\sum L = 185$

(a) Plot the means for each of the samples on the control chart. **(2)**

1. Mean = 18 2. Mean = 18.35 3. Mean = 18.5

(b) State what action, if any, should be taken. **(2)**

1. No action
2. Test again
3. Stop the process and reset.

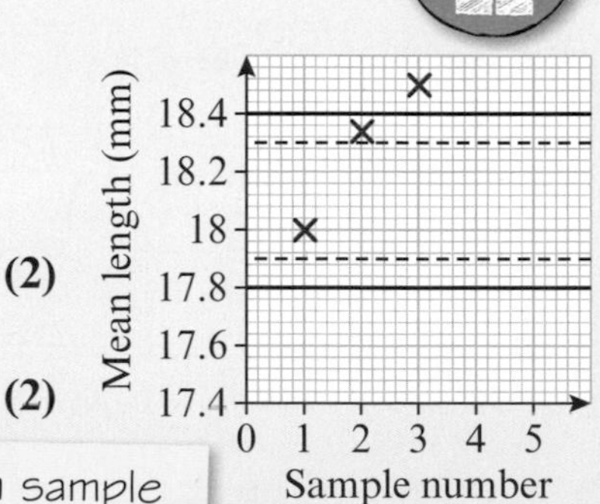

Always make it clear which sample you are discussing. For sample 2 the mean lies between the warning and the action limits.

Now try this

tier H

A machine fills bottles with liquid. 12 bottles with volume V ml of liquid in them are tested every 10 minutes. The distribution of the sample means is normal with $\mu = 330$ ml and $\sigma = 2.4$ ml.

(a) Work out warning and action limits for this sampling process. **(2)**

(b) One sample has $\sum V = 3398$. What action should be taken? **(2)**

Quality assurance and control charts 2

Quality control charts can also use the **median** or the **range** when monitoring the quality of goods. In each case the process is similar – samples are taken periodically and tested. Action, if any, is then taken on the basis of the test results.

Control charts and sample medians

The control chart for medians looks the same as that for means.

For each test the value of the median of the items in the sample, the **sample median**, is plotted.

- **Warning limits** are $\pm 2\sigma$ from the target median.
- **Action limits** are $\pm 3\sigma$ from the target median.

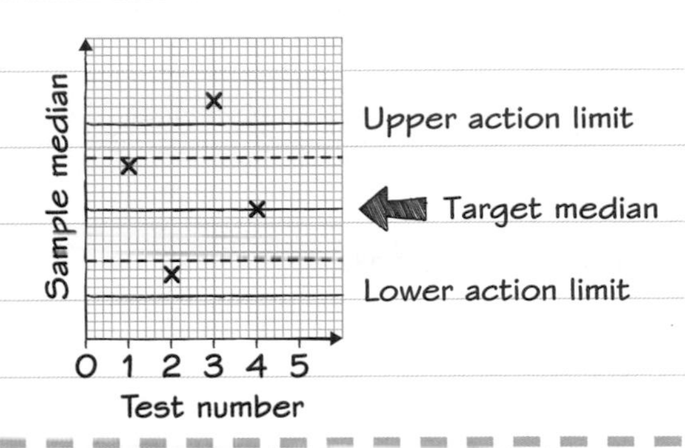

Control charts and sample ranges

As well as monitoring the mean or median value of samples from a production line, companies also monitor the variability of the items on the line.

It is possible, for example when producing biscuits, that the average weight is on target but the **variability** of the weights is unacceptable.

The **range** is used to measure the variability, as it is much quicker to find than the standard deviation.

The control chart for ranges looks the same as for medians and means.

A lower limit is sometimes shown, as **no** variation may mean that the testing equipment is not working.

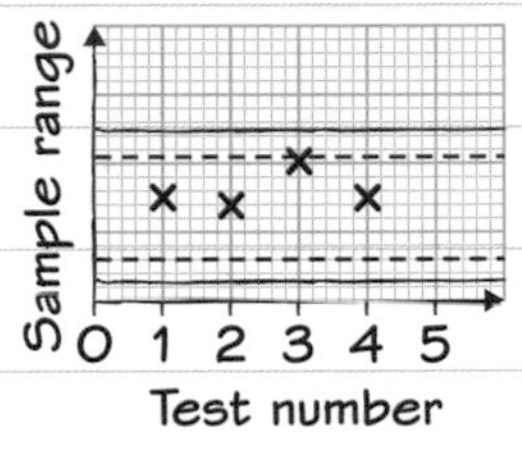

With some production lines the lower limits may be absent.

There is no simple relationship between the sample range and the normal distribution. The warning and action limits are based on the testing itself.

Worked example

tier H

The diagram shows a control chart for sample ranges in a biscuit factory. Some tests have already been done.

(a) Should the line have been stopped at any time? **(1)**

No, because all the ranges are within the warning limits.

For sample number 4, the weights in grams were

15.2 14.9 15.3 15.3 15.0 15.1 14.9 15.1

(b) Complete the control chart for sample 4. **(2)**

$15.3 - 14.9 = 0.4$

(c) State what action, if any, should be taken after sample 4. **(2)**

No action needed, as still within the warning limits.

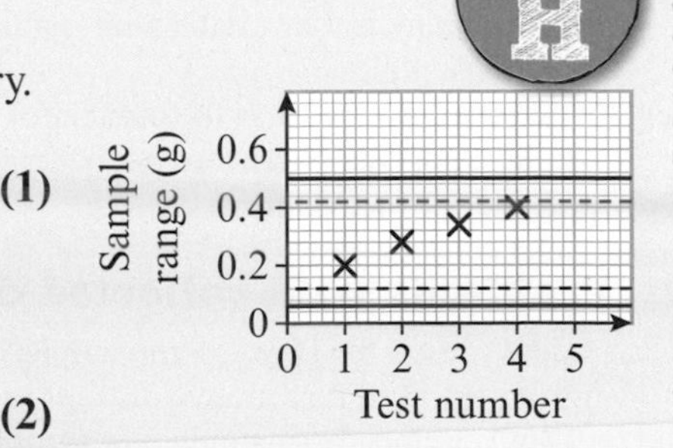

Although it is easy to do in your head, you should always show the calculation of the range.

Now try this

A control chart for medians is used to monitor a production line that makes cakes. The target weight is 125 g and $\sigma = 3.5$ g.

Work out the upper and lower warning and action limits for the control chart. **(3)**

Answers

COLLECTING DATA

1. Describing data

1 (a) D (b) C (c) B (d) A

2 Any suitable related data item, e.g. rainfall in month of holiday, cost of holiday, location such as city, inland or coast etc.

2. Primary and secondary data

(a) Secondary data, because it is obtained from the website.
(b) Data is reliable, because it is from a reliable Government source.
(c) Primary data, such as opinions of people who live in the postcode area, or secondary data, such as local newspaper reports or relevant social media in area.

3. Collecting data 1

1

Type of person	Tally	Frequency
Man		
Woman		
Boy		
Girl		

2

Type of precious stone	Tally	Frequency
Ruby	𝍸 III	8
Diamond	𝍸	5
Emerald	𝍸 I	6
Sapphire	𝍸	5

4. Collecting data 2

(a) Explanatory variable: type of formula,
Response variable: quality of dog's coat
(b) Advantage: easy to replicate; can control extraneous variables, e.g. amount of food given and environment.
Disadvantage: dogs will not be as relaxed out of their normal environment.

5. Collecting data 3

(a) Natural experiment
(b) An explanatory variable, in this case the number of visitors to the museum, is the variable that is affected by the response variable, (rainfall).
(c) An extraneous variable is the number of people who may be on holiday, or other activities happening over the summer which may attract visitors.

6. Problems with collected data

The student's result for Day 2 is anomalous because most students remembered 2 or more words more on Day 2 than they did on Day 1. Remembering only 8 words is an extreme outlier because all the other students remembered 15 or more. These results have been recorded accurately, so Brad may want to include it to give a full picture. The student may have been unwell or the words may have been unfamiliar.

7. Populations

(a) It is biased because his friends are likely to do similar things. Its size is too small.
(b) He should take a random sample of 30 or more using a school list of all the students in his year.

8. Grouping data

(a) Unsuitable. There are gaps between the intervals (e.g. cannot record 3.10).
(b) Unsuitable. The intervals overlap (e.g. 4.5 occurs in the first two intervals; 4.50 could be placed in 4.0 to 4.5 or 4.5 to 5.0 class interval).
(c) Suitable. There are no gaps or overlaps.

9. Random sampling

(a) Write all the numbers with 3 d.p. and multiply by 1000. The list becomes
583 958 196 811 680 43 326 374 416 6 334 719
Reject any numbers greater than 750.
Answer: 583, 196, 680, 43, 326, 374, 416, 6, 334, 719
(b) The sample is too small to be reliable. 10 out of 750 only represents $1\frac{1}{3}$% of the population.

10. Stratified sampling 1

27

11. Non-random sampling

1 (a) Select a number n from 1 to 4 at random. Then select the nth house, the $(n + 4)$th house, the $(n + 8)$th house, and so on.
(b) This takes one house from each block and in the same position in each block so is not fully random.
(c) Number all the houses in the street and select by using a list of random numbers.

2 (a) Because it is quick and cheap to do.
(b) Stop and test men until 12 have been interviewed. Stop and test women until 28 have been interviewed.

12. Stratified sampling 2

14

13. Petersen capture-recapture formula

(a) 520
(b) One of: that the marked bears are no more likely to be captured than the unmarked bears; that the marked bears are no more likely to die than the unmarked bears; that the population of marked bears does not change between the two samples.

14. Controlling extraneous variables 1

(a) The explanatory variable is the type of medication.
(b) The response variable is the patient heart rate.
(c) e.g. Patients could be affected by being in hospital and not sleep well. Researcher could consider giving patients heart monitors to use at home.

15. Controlling extraneous variables 2

(a) The purpose of the control group is to provide a comparison between the rose bushes in the control group and those in the test group. For example, the height of the rose bushes or the number of flowers they produce.
(b) The amount of sunlight the rose bushes receive during the day could affect their growth. The gardener would need to make sure they were all placed in a position where they received the same amount of sunlight.

16. Questionnaires and interviews 1

(a) The question is too open.
The response boxes allow only favourable responses.

(b) An improved question would be:
What did you think of the quality of the service you received today?
Very poor
Poor
Neither poor nor good
Good
Very good

17. Questionnaires and interviews 2

$520 + 480 = 1000$

$P(H) = \frac{1}{2}$

Estimated number of Heads $= \frac{1}{2} \times 1000 = 500$

Estimate for the number that ticked box A who were truthful $= 520 - 500 = 20$

Estimated proportion of people who had pretended to be someone else $= \frac{20}{1000 - 500} = \frac{20}{500} = 0.04$

18. Hypotheses

(a) For example: More people buy their clothes at large retail outlets than at the local shops or online.

(b) Ask people to record how they buy clothes by each method over 1 month.

(c)

Number of items of clothes bought this month	Tally	Frequency
Online		
Local shops		
Large retail outlets		

19. Designing investigations

(a) In an interview people might not give accurate answers if they are embarrassed or want to impress the interviewer. However, interviewers can be trained in asking sensitive questions.
In a questionnaire, people could answer anonymously or use a random response method, but if they are embarrassed they may not answer at all.

(b) There is a lot of data in one place so it is quick, however; some of the sources may be out of date or not reliable.

REPRESENTING DATA

20. Tables

(a) United States
(b) Germany, Italy and Japan
(c) Japan: $26.9 - 24.3 = 2.6\%$

21. Two-way tables

(a)

	School	Packed	Home	Total
Boys	10	4	3	17
Girls	7	4	2	13
Total	17	8	5	30

(b) 8

22. Pictograms

1

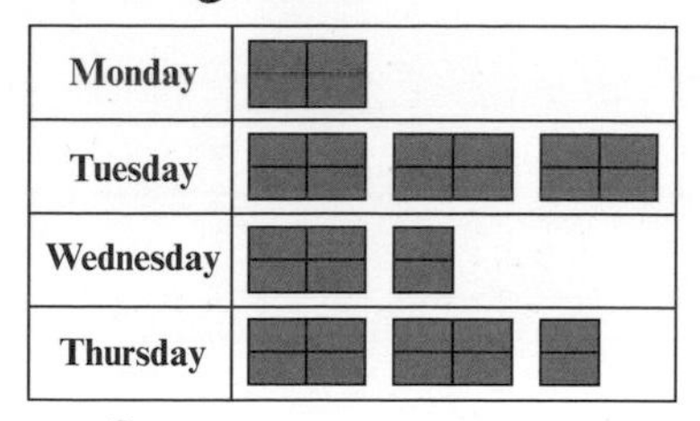

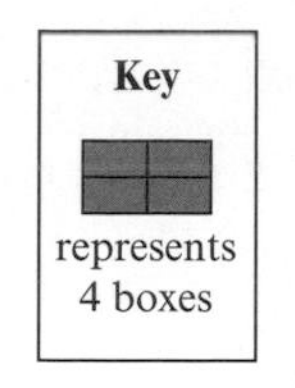

2 (a) ● represents 4 pets
(b) 11 dogs, 8 rabbits

23. Bar charts 1

1

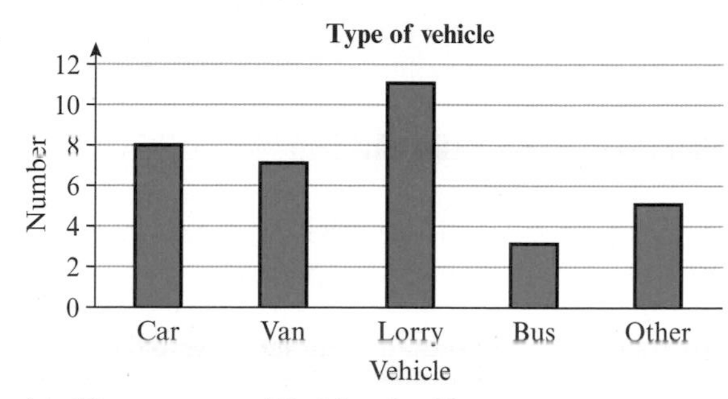

2 (a) 18 (b) $36 - 6 = 30$

24. Bar charts 2

1 (a)

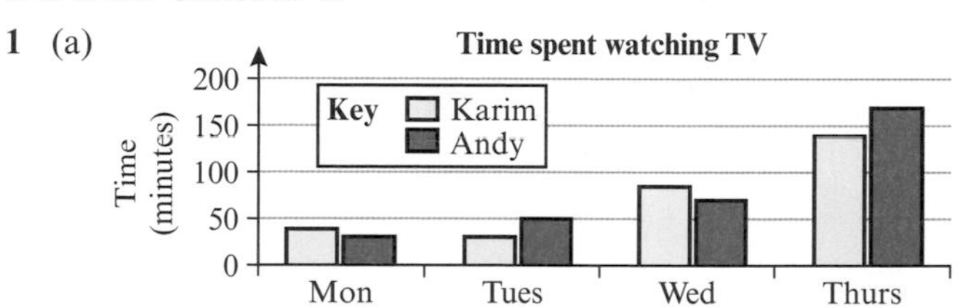

(b) The amount of time that Andy spent watching TV increased throughout the week.

25. Stem and leaf diagrams

1

10	4 5 8
11	0 2 3 9
12	0 1 2 2 7
13	1 2 3 7

Key: 10 | 4 represents 104 cm

2 (a)

Male babies		Female babies
	2	7 9
9 8 7 6 5	3	0 2 2 3 3 6 8
8 6 5 3	4	1
0	5	

Key: 6 | 4 = 4.6 kg 3 | 8 = 3.8 kg
or 6 | 3 | 2 represents 3.6 kg and 3.2 kg

(b) The male babies are heavier in general than the female babies.

26. Pie charts 1

(a) $\frac{72}{360} \times 2000 = 400$ kg
(b) 600 kg
(c) 144°

27. Pie charts 2

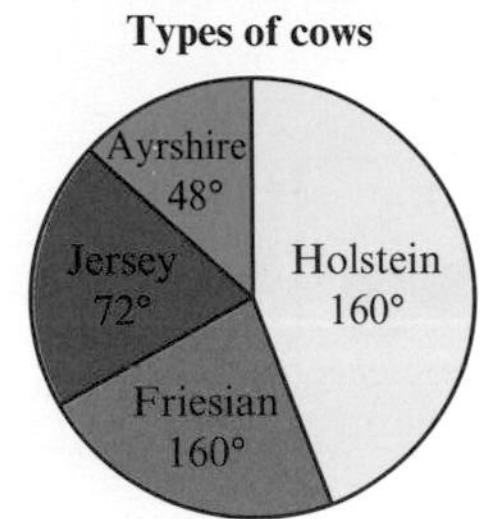

28. Comparative pie charts

(a) $\sqrt{\frac{450}{200}} \times 4 = 6\,\text{cm}$

(b) There are more of this type of weed in the moorland sample because it has a larger area.

29. Population pyramids

The proportion of people in the 75 and over age group is predicted to have a large increase by 2050 (about 8% in 2016 compared to nearly 15% in 2050). The proportion in the 60–74 age group is predicted to increase slightly and the rest of the age groups are predicted to show a slight decrease and the proportions are much closer together.

30. Choropleth maps

(a)

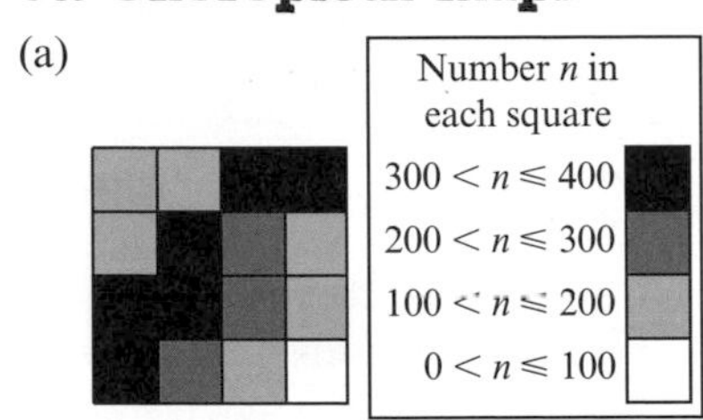

(b) The insects are distributed unevenly, with a band of higher insect density from bottom left to top right of the field.

31. Histograms and frequency polygons

1

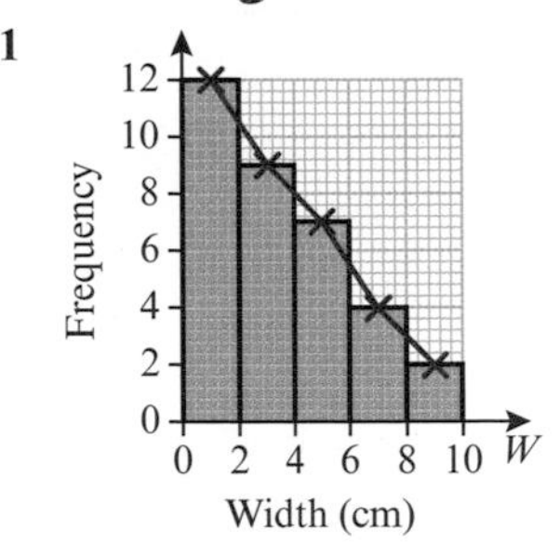

2 35

32. Cumulative frequency diagrams 1

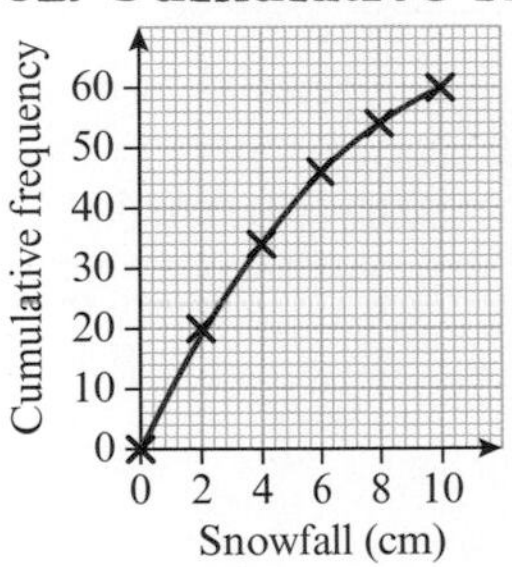

33. Cumulative frequency diagrams 2

1 (a) 44% (b) 40 years

2 (a)

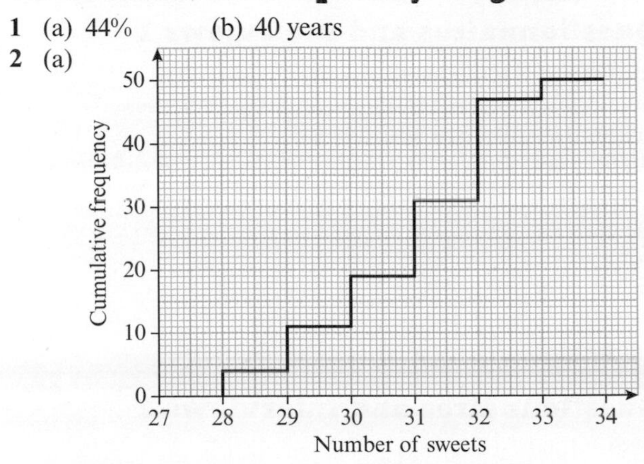

(b) 50

(c) 62%

34. The shape of a distribution

The distribution has positive skew.

35. Histograms with unequal class widths 1

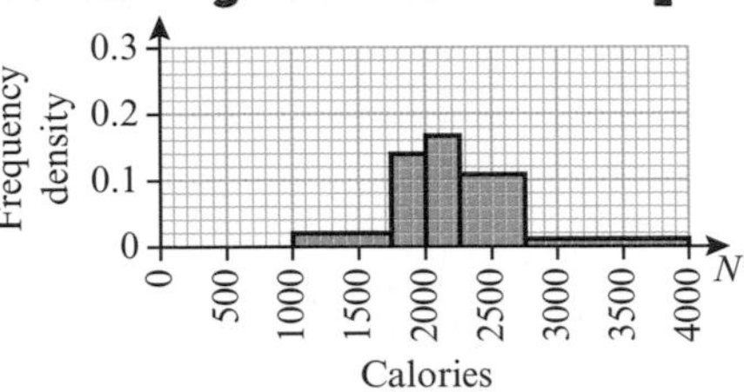

36. Histograms with unequal class widths 2

1 (a)

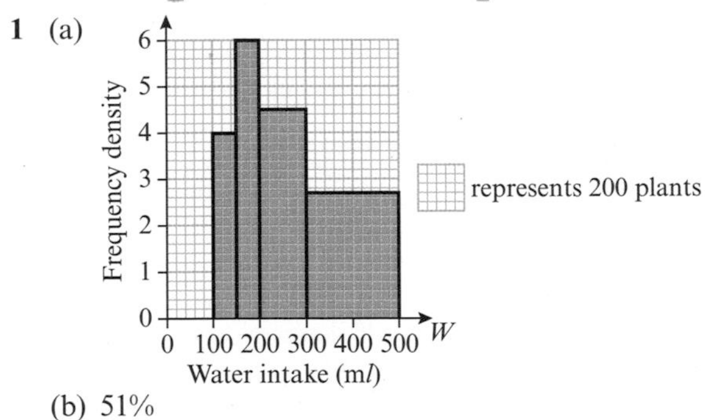

(b) 51%

37. Misleading diagrams

1 Vertical scale is not linear – goes 0, 100, 250 instead of 0, 100, 200.
Widths of bars are different – looks as if the number of males may be more than the number of females.
No vertical axis label – no idea what the graph is about (no title).

2 (a) Horizontal axis is not linear – sequence goes 50, 55, 65 instead of 50, 55, 60.
Vertical axis is frequency, not frequency density – class widths are unequal.
Bars should not be equal widths **or** frequency density has not been calculated.

(b)

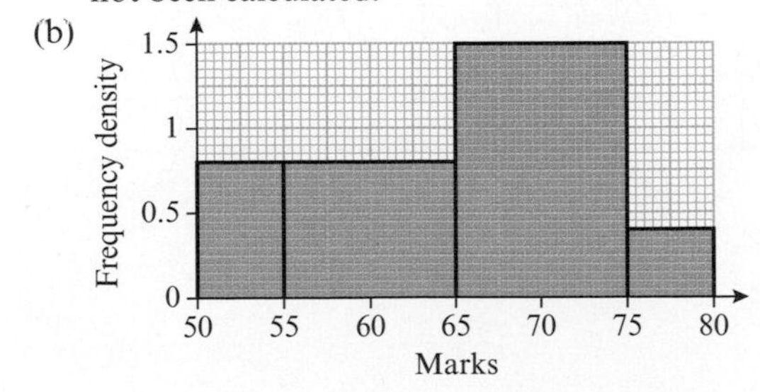

38. Choosing the right format

(a) A histogram is used for continuous data and the number of messages is discrete.
(b) Any one of: bar chart, vertical line graph or pie chart. Bar charts and line graphs are used to show trends or patterns, and to show totals; pie charts show proportion.

SUMMARISING DATA

39. Averages

(a) (i) 18 (ii) 18 (iii) 20.3
(b) Less than the mean, as 20 is less than 20.3.

40. Averages from frequency tables 1

(a) 24 (b) 24

41. Averages from frequency tables 2

2.4

42. Averages from grouped data 1

(a) (i) $0 \leqslant D < 5$ (ii) $10 \leqslant D < 15$
(b) 10.9

43. Averages from grouped data 2

(a) 347 m^2 (3 s.f.)
(b) The data is grouped so the actual size of each plot of wasteland is unknown.

44. Averages from grouped data 3

(a)

Height, h (cm)	Frequency (f)	Midpoint (x)	fx
$110 \leqslant h < 120$	6	115	690
$120 \leqslant h < 140$	22	130	2860
$140 \leqslant h < 155$	54	147.5	7965
$155 \leqslant h < 160$	4	157.5	630
Total	86		12145

Mean $= \frac{12145}{86} = 141.22$ cm (2 d.p.)

(b) 144.17 cm (2 d.p.)

45. Transforming data

(a) 489 000 (b) 513 000

46. Geometric mean

1 2.439
2 1.1 (1 d.p.)

47. Weighted mean

70%

48. Measures of dispersion for discrete data

(a) 8 (b) 3.75
(c) Advantage: it is easy to calculate.
Disadvantage it is affected by outliers.

49. Measures of dispersion for grouped data 1

(a)

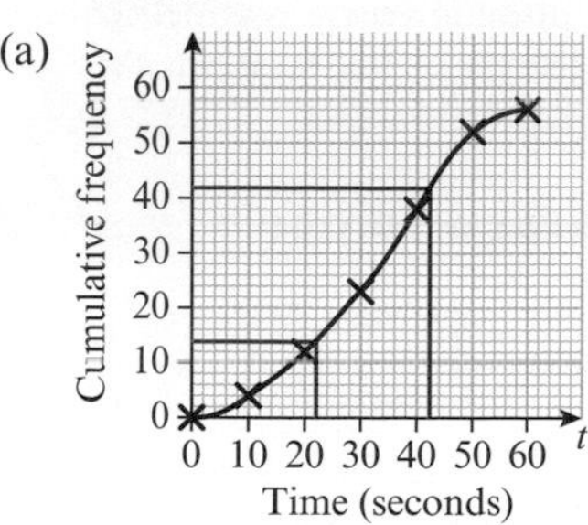

IQR = 20 seconds

(b) 19.61 seconds

50. Measures of dispersion for grouped data 2

(a) 17.8 (3 s.f.) (b) 9.4 (3 s.f.)

51. Standard deviation 1

Mean number of visits per day = 20, standard deviation = 8.9

52. Standard deviation 2

(a) Standard deviation = 4
(b) Mean will stay the same, standard deviation will go down.

53. Standard deviation 3

1 Mean $= \frac{1200}{30} = 40$

Variance $= \frac{51000}{30} - 1600 = 1700 - 1600 = 100$

Standard deviation = 10

2 (a) 148.5 cm (b) 12.19 cm
(c) The midpoint of each interval is used as an estimate for the value of each data item in that interval, and not the exact data value. So, the mean and standard deviation calculated from this estimate are also estimates.

54. Box plots

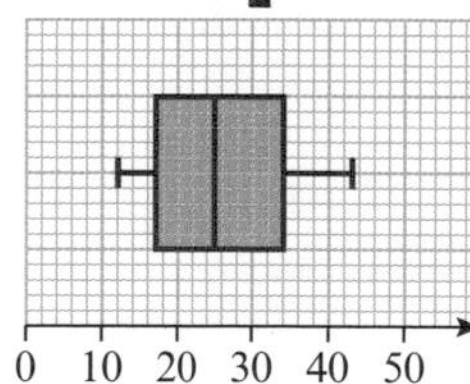

55. Outliers

1 (a) $Q_1 = 8, Q_3 = 17$ (b) There are no outliers.
2 4.6 is less than $9.05 - 3 \times 1.4 = 4.85$ so is an outlier.

56. Skewness

(a)

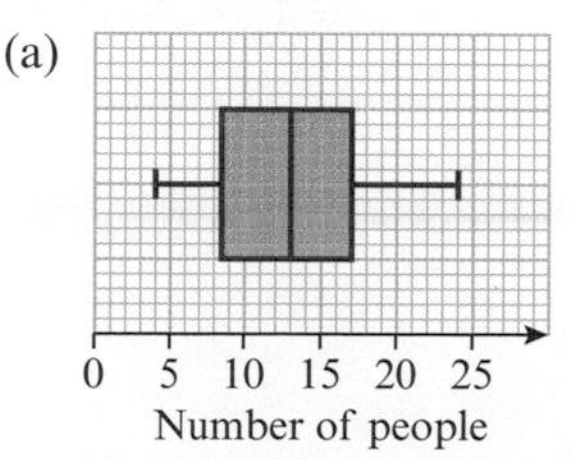

(b) The median is almost exactly in between the two quartiles so the distribution is symmetrical. There is no tendency for either a large attendance or a small attendance.
(c) Mean = 13.2 (3 s.f.)

(d) Median = 13, mode = 17. The mode is greater than the median but the median and mean agree to 2 s.f. which supports conclusion the distribution is nearly symmetrical.
(e) Skew = 0.09 which is weak positive skew. This supports the conclusion the distribution is nearly symmetrical.

57. Deciding which average to use

1 (a) 2
(b) 2.1 (1 d.p.)
(c) The mean will increase by 0.4. The median will not change.
2 Weighted mean, because there are different percentages of students for each range of text messages.

58. Comparing data sets

1 Group A: 0 1 2 3 4 5 6 6 7 8 8 9 10 12 13
Median = 6, $Q_1 = 3$, $Q_3 = 9$, IQR = 6
On average, the people in group A went to fewer shops. Group B visits are more clustered about the median.
2 The mean for the length of the American cars (498 cm) is greater than the mean length for the European cars, showing that the American cars are longer than the European cars on average. The standard deviation for the lengths of European cars (34.3) is lower than the standard deviation for the lengths of the American cars (50.8), showing that there is more variation in the lengths of the American cars.

59. Making estimates

1 (a) 617 (b) 2469
2 (a) 35% (b) 7.675 million

CORRELATION

60. Scatter diagrams

(a) It is a good diagram as rainfall and temperature are bivariate data.
(b) There is an association between the temperature in °C and the rainfall in mm. As the temperature increases the rainfall decreases.

61. Correlation

(a) Explanatory: age, response: weight. Positive correlation.
(b) Explanatory: length, response: time. Positive correlation.
(c) Explanatory: temperature, response: number of cups of hot tea drunk. Negative correlation.
(d) Neither, as there is unlikely to be any correlation.

62. Causal relationships

1 **A** and **C**
IQ is not dependent on shoe size. The weight of certain items may increase the cost, but light items such as precious stones could cost more than heavier items such as pallets of bricks.
2 Height and age are causally related up to the age of around 18 years. Height and performance in an IQ test can both be affected by age and show a spurious correlation.

63. Line of best fit

1 (a)

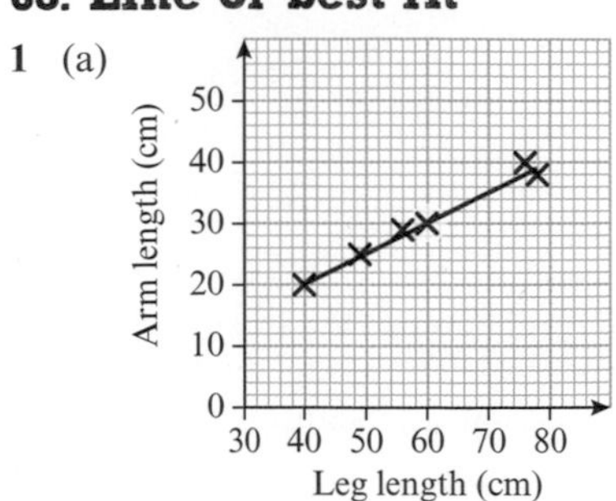

(b) 33 cm

2 (a)

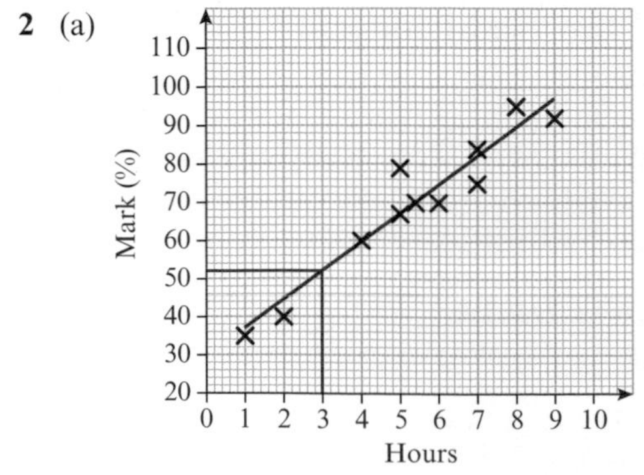

(b) Mean point (5.4, 69.9) (c) 52%

64. Interpolation and extrapolation

(a) (i) £10 500 (ii) Reliable as within the data
(b) (i) £1000 (ii) Unreliable as it is outside the data

65. The equation of a line of best fit 1

(a) $t = -1.4S + 12$
(b) The decrease in acceleration time for every litre increase in engine size.
(c) It predicts an acceleration time even when there is no engine present.

66. The equation of a line of best fit 2

(a)

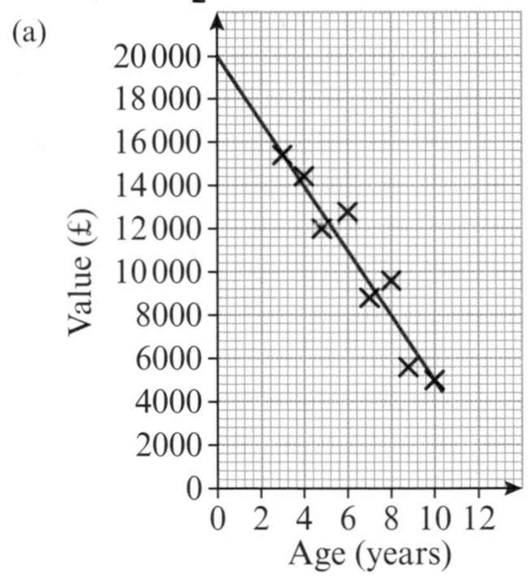

(b) The y-intercept is at £20 000. This is the cost of the car when new.
(c) The gradient is −1500, so the car decreases in value by £1500 every year.
(d) −£2500. This is not a possible value, which shows the regression line is not a reliable predictor above the range of given values.

67. Spearman's rank correlation coefficient

(a) The two judges' marks showed a strong positive correlation, so there was reasonable agreement on the rankings for the first dance for the two judges.
(b) Any two from: both correlation coefficients are positive and show agreement in the order of ranking for the two judges; there was less agreement between the judges on the ranks for the second dance; the correlation coefficient for the second dance shows weaker correlation.

68. Calculating Spearman's rank correlation coefficient

(a) 0.95
(b) There is a very strong positive correlation so the longer the bird the larger the weight.

69. Pearson's product moment correlation coefficient

(a) Perfect negative correlation; all points lie on regression line.
(b) Weak positive correlation; points are scattered around regression line.
(c) Strong negative correlation; points lie close to regression line.

TIME SERIES

70. Line graphs and time series

(a)

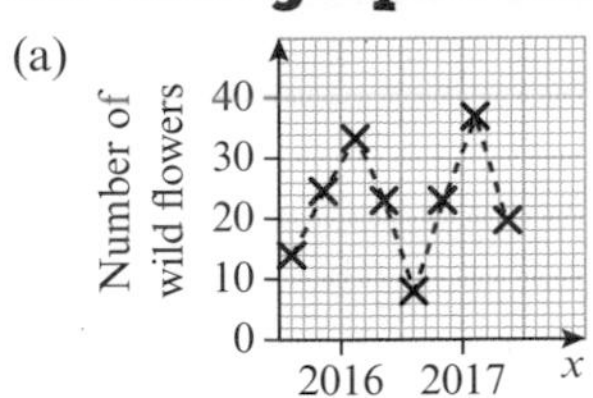

(b) Quarter 3 of 2017
(c) The most wild flowers were in Q_3.
The fewest wild flowers were in Q_1.

71. Trend lines

(a) and (b)

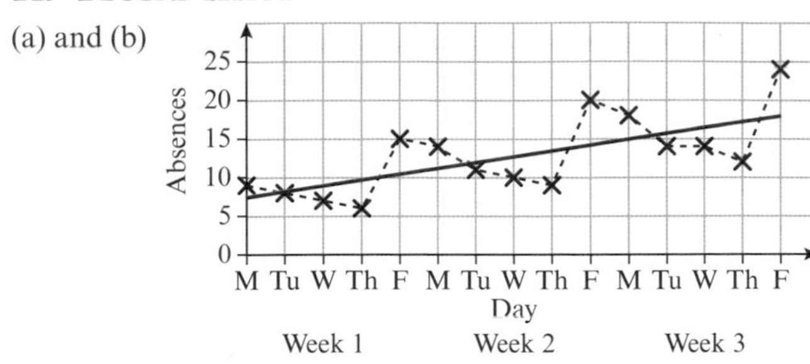

(c) The headteacher is right to be worried because ….
the trend for the number of absences on Friday is upwards (rising/increasing)
there are more absences on Friday than the other days of the week.
(Friday has the greatest variation above the trend line.)

72. Variations in a time series

The trend is approximately level: there is only a very slight increase over the two weeks.
The seasonal variation is by day. It is a seven-season cycle. Most tickets are sold on a Saturday. In each week Sunday, Monday, Friday and Saturday are higher than the trend value and Tuesday and Wednesday are below the trend value.

73. Moving averages 1

(a) 37.5, 39.5, 42.75, 43.75, 42.75
(b) The trend in the number of sales is upwards.

74. Moving averages 2

(a)–(c)

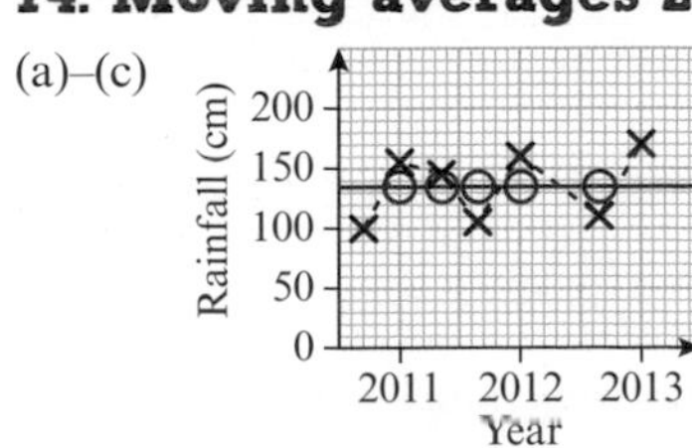

(d) The trend is level. The rainfall is neither increasing nor decreasing on average.

75. Seasonal variations

(a) and (b)

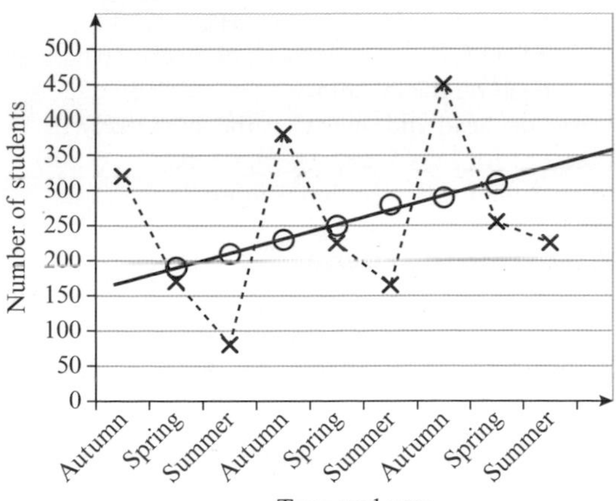

Autumn	Actual number of students	Trend (from line)	Seasonal variation
2015	320	170	+150
2016	380	230	+150
2017	450	290	+160

(c) (i) Mean seasonal variation for autumn
$$= \frac{+150 + 150 + 160}{3} = 153.33\ldots$$
(ii) Predicted value for autumn 2018 = 350 + 153
= 503 students

PROBABILITY

76. The meaning of probability 1

B at about 0.15, C at 0.5, A at 1.0 on a scale from 0 to 1.0

77. The meaning of probability 2

12 times

78. The meaning of probability 3

(a)

		Year group			
		Year 9	Year 10	Year 11	Total
Diet	Vegetarian	15	8	4	(27)
	Not vegetarian	19	22	7	48
	Total	(34)	(30)	11	75

(b) (i) Not vegetarian (ii) $\frac{30}{75}$ (iii) $\frac{4}{75}$

79. Experimental probability

1 (a) $\frac{110}{250}$ (b) $\frac{140}{250}$

2 (a) Because he rolls the dice the most times.

(b)

	Abel	Beth	Connor
Number of 2s	15	24	36
Number of rolls	60	90	150
Estimate of a probability of a 2	$\frac{15}{60} = 0.25$	$\frac{24}{90} = 0.27$	$\frac{36}{150} = 0.24$

(c) (i) Combine all three people's results.
(ii) $\frac{75}{300} = 0.25$

80. Using probability to assess risk

1 (a) 3.1 (1 d.p.) (b) 0.27 (2 d.p.)
2 (a) 2.64 (3 s.f.)
(b) 2.64 > 2 means the risk is over 2 times as high.
Tia is correct.

81. Sample space diagrams

1 (a) (1H, 2H), (1H, 2T), (1T, 2H), (1T, 2T) where (1H, 2H) means head on the 1p coin and head on the 2p coin.
(b) One head and one tail is the most likely.

2 (a) (A, B), (A, C), (B, A), (B, C), (C, A), (C, B)
(b) A man and a woman is more likely.

82. Venn diagrams

$x = 0.05$, $y = 0.02$, $z = 0.02$

83. Mutually exclusive and exhaustive events

(a) 0.28 (b) 0.4

84. The general addition law

$\frac{1}{5}$

85. Independent events

(a) 0.3 (b) $0.4 \times 0.25 = 0.1$
(c) $0.05 \times 0.05 = 0.0025$ (d) 0.005

86. Tree diagrams

(a)

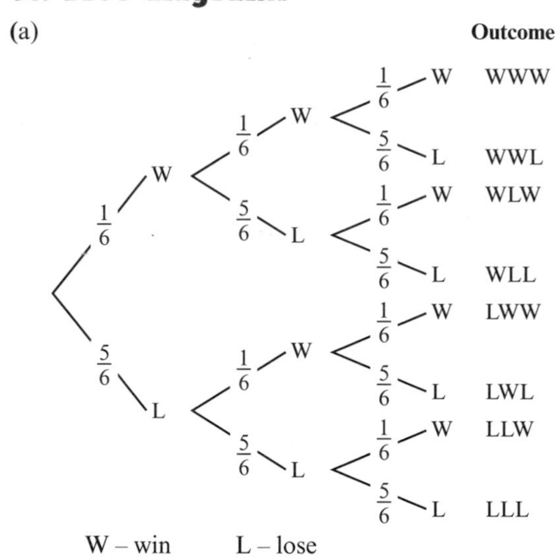

W – win L – lose

(b) $P(WWW) = \frac{1}{6} \times \frac{1}{6} \times \frac{1}{6} = \frac{1}{216}$

87. Conditional probability

$\frac{42}{90}$

88. The formula for conditional probability

(a)

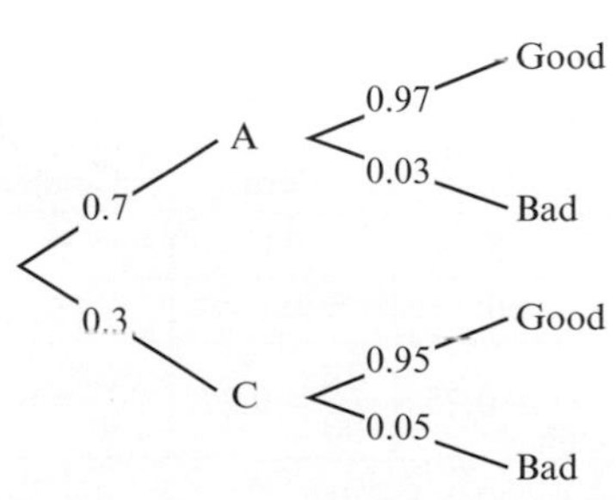

(b) (i) 0.7 (ii) 0.97 (iii) 0.679 (iv) 0.964

2 $P(A) \times P(B) = 0.4 \times 0.5 = 0.2 = P(A \text{ and } B)$ so A and B are independent.

INDEX NUMBERS

89. Index numbers

(a) 186.1 (b) £5.30

90. RPI and CPI

Price in 2017 as a percentage of the price in 1987

$= \frac{223\,257}{40\,000} = 558.1$ (1 d.p.)

This is more than 2 times the RPI.

91. GDP

1 (a) Increase: USA 0.5% Q2; Decrease: Japan 0.5% Q4
(b) UK: maximum percentage change in GDP was 0.2%

92. Weighted index numbers

Weighted index number = 106.4 (1 d.p.)

93. Chain base index numbers

(a) 102 and 102
(b) The chain base means that the base values for the two years are different.

94. Crude rates

43.7 (1 d.p.) per 1000 of the population of working age.

95. Standardised rates

Town A – standardised death rate $= \frac{60.5}{1000} \times 152 = 9.196$ means just over 9 deaths per 1000
Town B – standardised death rate $= \frac{72.3}{1000} \times 141 = 10.1943$ means just over 10 deaths per 1000
No, Abdul is not correct, $10.1943 - 9.196 = 0.9983$ so difference is just under 1 per 1000 of the > 65 population.

PROBABILITY DISTRIBUTIONS

96. Binomial distributions 1

(a) 0.32 (b) 0.64

97. Binomial distributions 2

(a) 0.000 075 9375 (b) 0.0022275
(c) $500 \times 0.15 = 75$ SUVs

98. Normal distributions

(a) 2.8 kg (b) 1.8 kg and 3.0 kg
(c)

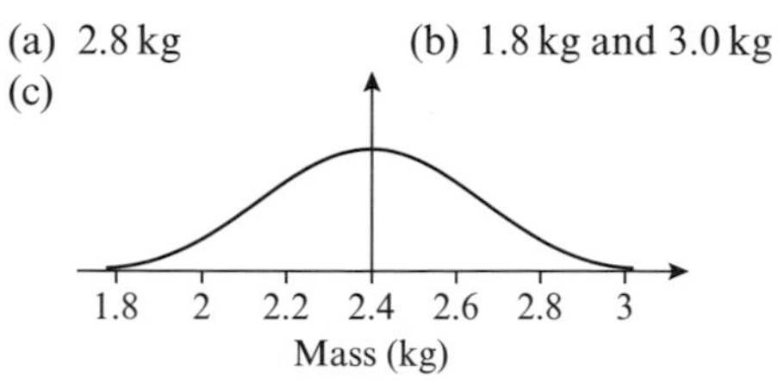

(d) 16%

99. Standardised scores

1 Aptitude 0.3, skills 1.125
2 (a) −0.5 (b) 64

100. Quality assurance and control charts 1

(a) Warning limits: 325.2 ml and 334.8 ml
Action limits: 322.8 ml and 337.2 ml
(b) The mean is 283 ml so the production line should be stopped.

101. Quality assurance and control charts 2

Warning limits: 118 g and 132 g
Action limits: 114.5 g and 135.5 g

Notes

Published by Pearson Education Limited, 80 Strand, London, WC2R 0RL.

www.pearsonschoolsandfecolleges.co.uk

Copies of official specifications for all Pearson qualifications may be found on the website:
qualifications.pearson.com

Typeset and illustrated by Tech-Set Ltd, Gateshead
Produced by ProjectOne
Cover illustration by Miriam Sturdee

The right of Su Nicholson to be identified as author of this work has been asserted by her in accordance with the Copyright, Designs and Patents Act 1988.

First published 2018

21 20 19 18
10 9 8 7 6 5 4 3 2 1

British Library Cataloguing in Publication Data
A catalogue record for this book is available from the British Library

ISBN 978 1 292 19162 1

Printed in Slovakia by Neografia.

Acknowledgements
Content by Rob Summerson is included.

The author and publisher would like to thank the following individuals and organisations for permission to reproduce copyright material:

p20 Crown Copyright: House of Commons Library Briefing Paper 23 August 2017 Contains Parliamentary information licensed under the Open Parliament Licence v3.0; HM Land Registry, Registers of Scotland, Land and Property Services Northern Ireland and Office for National Statistics http://landregistry.data.gov.uk/app/ukhpi . Contains Parliamentary information licensed under the Open Parliament Licence v3.0. **p20 The World Bank Group:** Population ages 65 and above (% of total) data.worldbank.org/indicator/SP.POP.65UP.TO.ZS **p29 Crown Copyright:** Office for National Statistics www.ons.gov.uk/peoplepopulationandcommunity/populationandmigration/populationestimates/bulletins/annualmidyearpopulationestimates/mid2016 Contains public sector information licensed under the Open Government Licence v3.0. **p45 Crown Copyright:** Department of Digital, Culture, Media and Sport www.gov.uk/government/statistical-data-sets/museums-and-galleries-monthly-visits Contains public sector information licensed under the Open Government Licence v3.0 **p46 Crown Copyright:** Office for National Statistics www.ons.gov.uk/peoplepopulationandcommunity/populationandmigration/populationestimates/articles/overviewoftheukpopulation/july2017 Contains public sector information licensed under the Open Government Licence v3.0 **p59 Crown Copyright:** Office for National Statistics Contains public sector information licensed under the Open Government Licence v3.0 **p80** Publicly available data taken from www.ambulanceresponsetimes.co.uk **p90 Crown Copyright:** Office for National Statistics www.statista.com/statistics/281724/consumer-price-index-cpi-united-kingdom-uk-y-on-y/ contains public sector information licensed under the Open Government Licence v3.0 **p91 Crown Copyright:** Office for National Statistics www.ons.gov.uk/economy/grossdomesticproductgdp/timeseries/ihyr/ukea contains public sector information licensed under the Open Government Licence v3.0 www.ons.gov.uk/economy/grossdomesticproductgdp/bulletins/grossdomesticproductpreliminaryestimate/octobertodecember2017 contains public sector information licensed under the Open Government Licence v3.0 **p94 Crown Copyright:** Office for National Statistics www.ons.gov.uk/peoplepopulationandcommunity/birthsdeathsandmarriages/livebirths/datasets/birthsummarytables contains public sector information licensed under the Open Government Licence v3.0

Notes from the publisher
1. While the publishers have made every attempt to ensure that advice on the qualification and its assessment is accurate, the official specification and associated assessment guidance materials are the only authoritative source of information and should always be referred to for definitive guidance.

Pearson examiners have not contributed to any sections in this resource relevant to examination papers for which they have responsibility.

2. Pearson has robust editorial processes, including answer and fact checks, to ensure the accuracy of the content in this publication, and every effort is made to ensure this publication is free of errors. We are, however, only human, and occasionally errors do occur. Pearson is not liable for any misunderstandings that arise as a result of errors in this publication, but it is our priority to ensure that the content is accurate. If you spot an error, please do contact us at resourcescorrections@pearson.com so we can make sure it is corrected.